Labor and the Environment

Labor and the Environment

An Analysis of and Annotated Bibliography on Workplace Environmental Quality in the United States

Compiled by

Frederick H. Buttel,
Charles C. Geisler,
and
Irving W. Wiswall

Greenwood Press
Westport, Connecticut • London, England

Library of Congress Cataloging in Publication Data

Buttel, Frederick H.
  Labor and the environment.

  Bibliography: p.
  Includes index.
  1. Environmental policy—United States—Bibliography.
  2. Environmental policy—Cost effectiveness—Bibliography.
  3. Work environment—United States—Bibliography.
  4. Industrial hygiene—United States—Bibliography.
  I. Geisler, Charles C.   II. Wiswall, Irving W.
  III. Title.
  Z5863.P6B87   1984   [HC110.E5]   016.331   83-22575
  ISBN 0-313-23935-5 (lib bdg.)

Library of Congress Catalog Card Number: 83-22575
ISBN: 0-313-23935-5

First published in 1984

Greenwood Press
A division of Congressional Information Service, Inc.
88 Post Road West, Westport, Connecticut 06881

Printed in the United States of America

10 9 8 7 6 5 4 3 2 1

Contents

Preface

As this work goes to press, there is diverse evidence of environmental revivalism in the United States. Central to this revivalism, and to the motivation behind this volume, is the emerging alliance between labor and environmentalists. Labor-environmental networks and coalitions have formed to lobby for full employment, to oppose highway and import legislation vexing labor, and to withstand federal efforts to eviscerate health and safety regulation in the workplace. Among the most uncompromising of the clean-environment proponents, according to a Harris poll commissioned by Business Week early in 1983, are skilled laborers and union members.

In 1980 such developments were perhaps unforeseen. Single-issue politics and the narrowness it implied were the order of the day. The environmental euphoria of Earth Day, a decade before, was on the wane. Environmental causes were criticized as elitist--anything but fanfare for the common man. And a president was elected who, along with his cabinet appointees, construed his election victory as a mandate to deregulate the environment in the workplace and beyond.

Pollster Louis Harris attributes the environmental revivalism, evident in the 1983 national survey mentioned, to public concern over the perceived hostility of Interior Secretary James G. Watt and then Environmental Protection Agency Administrator Anne M. Burford to the laws they are charged with enforcing. Such an accounting portrays public concern too narrowly, if not too personally. Public sentiment on behalf of the environment has strong historical antecedents in the United States. Even in 1980, for example, despite political winds to the contrary, that sentiment was clearly evident in a survey by the Council on Environmental Quality, in which the public expressed a strong commitment to improved environmental quality "at all costs" (Congressional Quarterly, 1981:6). Moreover, Harris's explanation is inattentive to the growing detente between labor and environmental interests evident in such organizations as Environmentalists for Full Employment, the Workers' Institute for Safety and Health, and the AFL-CIO coalition with environmentalists, the Labor-Environmental Network.

There is, in other words, a new energy pulsating in the country on behalf of environmental protection. This new energy is the second wind of an older, more primal energy. Its elitist content is dissolving as environmentalists, labor leaders, and occasionally business envoys bury old hatchets and explore new partnerships. Underlying this coalition of interests is the spreading perception that distinctions between the "natural" and workplace environment are in many ways artificial. In years to come, more and more jobs will be located in nontraditional workplaces, blurring simple town-country boundaries of the past.

Two bodies of literature are examined in this volume. The first is unusual in that it collects from many quarters the complex account of cooperation between labor and environmental interests. Veterans from both movements will appreciate and perhaps be surprised by the vastness of this literature and the kindred interests of two groups commonly seen as antagonistic. If it is true that the labor movement is bringing new vitality to the contemporary environmental movement, it is also true that environmentalists have substantive things to offer working people. As the second body of literature makes clear, environmentalists have had more than a decade of frontline experience with benefit-cost analysis-- experience of obvious use to labor in responding to BCA challenges by the business community and the Reagan administration to such things as the Occupational Health and Safety Act of 1970. The sources assembled will assist labor in grasping the strengths, weaknesses, biases, and subtleties of benefit-cost analysis as they apply to workplace environments.

The first set of annotations, then, assembled a recent segment of social history in which two major movements are intersecting. The second expands upon a subtheme within this larger literature which will, if our intuition is correct, figure prominently in the labor-environmental reciprocity and common cause of the future. Both sets of annotations are introduced with essays intended to clarify, in greater depth and detail, the logic of extending the literature summarized herein to new constituencies. Index references at the end of the book refer to annotation entry numbers rather than to page numbers.

The authors wish to thank several individuals for their determined and devoted help in preparing this manuscript. Julian McCaull and Helene Moran Vigorita generously gave of their professional time and talent in an editing capacity. Holly Ladra and Francene Signor typed early drafts of the manuscript and, together with Mary Lamberts, patiently helped with corrections. Many others shared source material for which we are grateful.

F.H.B.

C.C.G.

I.W.W.

Labor and the Environment

Part I

Labor's Stake in Environmental Quality and Energy Policy

FREDERICK H. BUTTEL

The environmental movement of the past two decades has brought both environmental and energy issues to the fore of political and scholarly debate. Labor has a huge stake in both environmental and energy policies. Consequently, relations between environmentalists and labor--both union and nonunion[1]--have become a major academic and political concern.

Energy problems have a preeminent institutional impact on labor. Likewise, efforts at environmental reform inevitably raise questions of distribution and equity that are of concern to labor--questions of who benefits from environmental protections and who pays for them. Answers from the U.S. labor movement have been largely negative towards environmental reforms, and the pattern of economic stagflation of the past ten years has served to intensify the response (see, for example, Wolfe 1981).[2]

Not surprisingly, the extended period of continuous inflation and economic stagflation has generated tremendous interest in identifying the origins of and solutions for these economic problems. Public and private expenditures on environmental protection have frequently been identified as major culprits in reducing the supply of credit, exacerbating inflation, and raising production costs, thereby constraining investment, employment, and the creation of jobs. In addition, there has been concern that the activities of environmentalists and public officials have crippled the ability of the energy-producing sectors to provide ample amounts of inexpensive energy for the production and consumption sectors of the economy, thereby causing inflation, dampening productivity gains, reducing the level of economic activity, and increasing unemployment.

Amid concerns that aggressive environmental and energy conservation policies might be inconsistent with the interests of American workers, a body of scholarship has emerged which suggests that these adverse impacts may have been exaggerated and that environmental protection and energy conservation may, on balance, be in the long-term interest of the U.S. labor movement. Many such studies, for example, have found that environmental protection expenditures have had little adverse effect on the economy as a whole while at the same time creating thousands of jobs in the pollution-control sector of the economy. Likewise, several studies have concluded that energy policies premised on energy conservation and the development of alternative, "soft"[3] energy sources would be accompanied by very substantial increases in employment. Thus there has been a growing debate as to whether organized labor and other American workers should support or oppose environmental and energy conservation initiatives.

This essay provides a broad overview of the academic and policy issues that surround labor's stake in environmental quality and energy policy. First, it presents these issues in the context of the emergence of the American environmental movement and the environmental conflicts that have developed as a result of that movement. The second part of the essay examines the major areas of research that have received attention with regard to labor and the environment.

It will become apparent that on the basis of our evaluation and interpretation of this literature, we are pursuaded that American workers, on balance, have not been and are not likely to be affected adversely by continued attention to environmental protection and, moreover, that public policies of energy conservation and development of renewable energy sources would be greatly beneficial for the U.S. labor movement. Many of the studies in the annotated bibliography which follows support these views. However, several studies report data contrary to our conclusion, and we have attempted to be as inclusive as possible in enumerating scholarly investigations into the relationship between environmental and energy policy and labor. Thus, the annotated bibliography is intended to be a research and policy resource that will be of general interest to those concerned with labor-environment issues and is not intended to be a vehicle for documenting the conclusions which we, nonetheless, feel are warranted.

The Context: The Environmental Movement and
Environmental Politics in the 1960s and 1970s

The modern environmental movement had significant roots in the "old-line" conservation movement that blossomed during the Progressive Era at the turn of the century (Harry 1974; Humphrey and Buttel 1982). Nevertheless, 1960s' environmentalism represented a fundamental shift in the nature and scope of environmental politics. The environmental movement greatly expanded the range of environmental problems that were of concern to both leadership and membership—shifting, generally speaking, from largely rural and aesthetic issues such as wilderness preservation and endangered species protection to more urban or "ecological" issues such as air and water pollution, toxic waste disposal, workplace contamination, and energy conservation. This shift in the issues of concern to environmentalists represented an expansion of the movement's agenda from recreational and aesthetic issues—which, not surprisingly, were of direct and immediate interest to traditional conservationists as outdoor recreationists and naturalists—to a set of issues wherein environmentalists did not necessarily have a direct material interest. For example, suspending construction of a power plant in New York State and securing government funding to decontaminate a chemical waste dump in North Carolina are actions which benefit environmentalists only as citizens, not as outdoor recreationists.

This expanded scope of the environmental movement was made possible by a unique sociopolitical phenomenon: the massive growth in public attention to and support for environmental initiatives during the late 1960s and early 1970s. While growth in environmental awareness and sentiment was not a phenomenon unique to the United States—there was a parallel mobilization of environmentalism in most other advanced industrial societies at the same time or shortly thereafter—the American pattern of environmental mobilization had a special character.

In the United States, environmentalism emerged within the context of two other then-prominent mass movements, the civil rights and antiwar movements, and this historical configuration heavily shaped the political and ideological thrust of the environmental movement in several respects. First, the largely college campus-based civil rights and antiwar movements were encountering impasses in 1969 and 1970; environmentalism became appealing because it seemed to offer a new frontier for youthful activism. Second, the most widely publicized "environmental disasters" which focused public attention on ecological issues—particularly the Santa Barbara oil spill and the delaterious impacts of pesticides on wildlife—tended to have corporate "villains." Thus, the transfer of youthful civil rights and antiwar activism environmental issues occurred quite rapidly. Third, this shift toward environmental activism led to a crucial role of college campuses in the early phase of the environmental movement. The dominant thrust of the early years of growth of mass environmentalism was campus based, which proved to be both a strength and a weakness for the movement in the years that followed.

Perhaps the key influence of the campus-based character of early environmentalism was the ability of idealistic collegiate eco-activists to connect a variety of natural resource and ecological phenomena under the broad rubric of "ecology." The general distrust of the late 1960s' cohort of college students toward public and private authority gave environmentalism a unique flavor: uncompromising, confrontational, and strongly oriented toward participatory and democratic strategies. Much of this sentiment may now be viewed as idealistic and perhaps even impractical with regard to the realities of American politics. Nevertheless, the campus rooting of environmentalism had a major influence on movement ideology and tactics; it is unlikely that the same sociopolitical coloration of the environmental movement would have emerged if the activity had been largely under the auspices of the old-line conservation organizations such as the Sierra Club, the Isaac Walton League, and the Wilderness Society.

The campus-based nature of the environmental movement quickly faded, however, because of the transitoriness of college students. Youthful eco-activists left college campuses in the early 1970s, which often led to the demise of the ecology groups that had been formed in 1969 and 1970. The locus of environmental activism increasingly shifted toward the established conservation organizations, with the Sierra Club, its schismatic offspring, Friends of the Earth, and several new but related groups such as the Natural Resources Defense Council playing the key roles in environmental activity.

The early environmental movement registered some rather rapid and major successes, highlighted by the passage of the National Environmental Policy Act of 1969 by a unanimous voice vote of the congressional branches. These early and stunning victories were due in part to the virtual "motherhood" character of ecology. It is also likely that politicians weary of campus disturbances were relieved to see campus activism siphoned off into what appeared to be a less threatening political agenda.

4 Labor and the Environment

An Initial Labor-Management Alliance

The motherhood aura of environmentalism faded when it became apparent that environmental reform and protection involved significant costs that eventually must be borne by some segment of society. Corporations (and, secondarily, the federal government) were the major targets of environmental initiatives, and corporate organizations were the first to take the offensive against environmentally inspired curbs on their decision-making prerogatives. It is significant, however, that corporations preferred not to become the sole members of the antienvironmental countermovement in an era in which ecology still had lingering elements of motherhood status.[4] Morrison (1973) and Sills (1975) noted that corporations attempted, often with considerable success, to enlist their workers as allies in the struggle against environmentalists. Corporations and certain segments of labor began to assume a united posture that environmental protection—if such a policy increased the costs of industrial production, reduced corporate profitability, and led to plant closings or layoffs—was not in the best interest of the majority of Americans.

In certain respects, this posture proved to have a significant thread of validity. Early simulations of environmental control policies by researchers such as Krieger (1970) indicated that environmental protection tended to have regressive socioeconomic impacts. Likewise, pioneering studies by Schnaiberg (1975) and others indicated that energy price increases, used as a mechanism to curb energy demand, tended to have disproportionately detrimental impacts on the working class and the poor.[5]

In a more general sense, the "post-motherhood" stage of environmentalism came to be dominated by polarization over "economic growth versus the environment." Environmentalists and their allies increasingly came to question the desirability of economic growth, and many scholars and activists (for example, Daly 1973, 1977) assumed an emphatic antigrowth posture, arguing that a zero-growth or stationary-state economy was preferable to the historical trajectory of economic expansion and environmental destruction that had prevailed in the United States in the post-World War II period.

The polarized issue of economic growth versus the environment subsided rapidly toward the end of the 1970s as economic stagflation strongly undermined support for environmental programs. Environmental groups began to back off from economic growth/environmental protection trade-off rhetoric, because the search for solutions to U.S. economic problems was leading to increasing assaults on existing environmental legislation. This rearguard action of environmentalists—which has continued in large measure to the present, was accompanied by a search for alternatives to the polarity of economic growth versus the environment. The search was for mechanisms by which environmental goals could be accomplished without exacerbating the economic crisis.

The tentative, and not universally accepted, solution to this dilemma has been "appropriate" or "soft" technology. Inspired by the late E. F. Schumacher's Small is Beautiful (1973), the perspective of appropriate technology—that energy and resource conservation is compatible with economic viability and expansion of employment—has become increasingly integrated into the ideological expressions and political agendas of environmentalists. It should be noted that the traditional core of the longstanding environmental groups has not fully embraced appropriate or soft technology. There remain a number of questions about the technical feasibility of many appropriate technologies, and many conservative members of the environmental groups are uncomfortable with the overt—often violence prone—antinuclear tendencies of appropriate technology groups. Nevertheless, the environmental movement has rapidly evolved through a number of stages and is increasingly finding "appropriate technology" to be a useful rhetorical device for generating coalitions between environmentalists and other groups such as labor.

The Reagan Administration and the Environment

The urgency of addressing public and private assaults on environmental policy has greatly increased with the assumption of power by the Reagan administration. The Reagan administration has taken an overtly antienvironmental posture, arguing that relaxation of automobile emission standards, of Occupational Health and Safety Administration (OSHA) regulations, of air and water quality statutes, and of other environmental laws and regulations must be undertaken to restore economic dynamism and growth. Moreover, the Reagan administration has been clearly supportive of the traditional "hard" energy sectors, especially nuclear power and domestic petroleum, and has virtually scuttled federally funded research on alternative, renewable, "soft" energy sources.

A significant statistic in this regard is that the Reagan administration was not elected over the vociferous objection of American workers. Various polls suggest that from 45 to 48 percent of AFL-CIO members voted for Ronald Reagan in 1980. While these data do not indicate that Reagan is "labor's president," the fact that Reagan had significant support among the members of organized labor indicates that the 1980s will be a crucial decade for the environmental movement in relation to the American labor movement.

One possible interpretation of the surprising performance of the Reagan-Bush ticket among organized labor is that labor union members tend to be antienvironmental and that Reagan's overtly antienviromental posture was a significant lever for attracting union votes in the general election. Such a view, however, cannot be substantiated from empirical research on environmental attitudes. Studies by Mitchell (1979) and Buttel and Flinn (1978), for example, demonstrate that occupation and union membership are not related to environmental attitudes among the general public. Thus, Reagan's appeal was nourished by economic roots rather than by a reservoir of antienvironmental sentiment among American workers.

Another recent development of major importance in the relationship between environmentalism and the labor movement was the U.S. Supreme Court ruling that OSHA regulations must be based on cost-benefit studies in which the standards set for workplace hazards and contaminants must yield benefits greater than the cost of implementing technological controls. This procedure is a marked departure from earlier premises that the workplace should be safe regardless of cost. Because of the importance of cost-benefit analysis to environmental dangers such as workplace contaminants and safety threats, a separate section of this annotated bibliography, along with a corresponding introductory essay, is devoted to such issues.

As is apparent from the foregoing discussion of the recent context surrounding the relationships between environmentalism and labor, the labor and environmental movements have exhibited a complex pattern of cooperation and conflict. This pattern reflects an underlying paradox in the interests of labor vis-a-vis environmental quality. On one hand, working-class families tend to experience a greater degree of exposure to pollution in both the workplace and in residential neighborhoods than do the middle-class and propertied segments of society. On the other hand, American workers are subject to insecurities in their employment status and are vulnerable to "environmental blackmail"--that is, to threats by their employers that they must reduce employment as a result of environmental protection measures. Hence, despite the fact that environmental improvements would be of long-term benefit to labor, the short-term vulnerabilities and anxieties among workers have created an ambivalent relationship between labor and environmentalists.

Environmentalism and Elitism

Many of the bibliographic items annotated below provide significant and hopeful evidence that progress is being made to minimize the rhetorical cleavages and conflicts of interest that often divide environmentalists and labor. However, the environmental movement has for some time been plagued by a lingering image--if not, in certain instances, a clear reality--that environmentalism is an elitist ethic for advancing the narrow interests of middle-class and affluent citizens. The accusation of elitism emerged shortly after the rise of the mass environmental movement--justifiably so, given the tendency for environmental groups to pursue their agendas without regard to the impacts on other members of society. Neuhaus's In Defense of People (1973) was perhaps the most dramatic early example of the argument that environmentalism was at root a superficial reflection of the interests of middle-class intellectuals and that the strategies pursued by the movement were adverse to the interests of the American working class.

As Morrison and Dunlap (1980) noted, the accusation of elitism must be separated into three distinct aspects of class bias. The first, compositional elitism, refers to whether environmentalists represent a privileged stratum of the population. All studies of organized environmentalists have tended toward a common conclusion: environmentalists, compared to the larger population, are considerably better educated, more likely to have white-collar jobs, and more likely to have high incomes (see, for example, Humphrey and Buttel 1982: Chapter 5). This finding, however, is quite similar to those reported for the composition of the membership of most voluntary social movement organizations. Most such organizations, including the civil rights and other movements aimed at benefitting the poor, are primarily composed of middle-class, well educated persons, generally because these persons have the educational and organizational skills crucial in securing funding for movement organizations (McCarthy and Zald 1973). Also, as suggested earlier, public attitudes toward environmentalism are not characterized by a strong polarization of opinion with regard to occupation or income (although there is a clear tendency for the well educated and the young to be more proenvironmental than the less well educated and the elderly). Thus, the available data on the extent of "compositional elitism" in the environmental movement do not unambiguously support the accusation that environmentalism is inherently elitist.

Ideological elitism--the possible tendency for environmentalists to advocate the distribution of benefits from environmental protection to a narrow stratum of environmentalists while paying little attention to the allocation of costs, especially among the poor--is an equally ambiguous issue. Many environmentalists, typified by Garrett Hardin (1974), have indeed advocated policies which are overtly hostile toward the poor in this country and abroad. Nevertheless, it is equally the case that other leading figures in the environmental movement, notably Barry Commoner, have severely criticized the notions of Hardin and others who advocate a hostile posture toward labor and the poor.

More generally, the environmental movement has tended to reflect a reformist liberal position which has been most compatible with the ideology of the Democratic Party--the national political party historically most concerned with the problems of labor and the poor (Dunlap and Allen 1976). Likewise, proenvironmental persons in the general public tend to be politically liberal and more likely than environmental opponents to be concerned with the problems of low-income and working-class people. Yet these otherwise favorable predispositions of the environmentally concerned toward the interests of labor do not imply that environmentalists shed narrow self-interest when their recreational and residential amenities are at stake, even though their actions at the ballot box, or their responses to surveys, might be sympathetic with the problems of the poor. On balance, it is apparent that environmentalists have had a mixed record in uniting their concerns with the concerns of the labor movement. However, recent evidence suggests greater attention to these matters as environmentalists search for allies to prevent a disastrous onslaught on their legislative and legal victories of the 1970s.

The final aspect of elitism involves the assertion that environmental protection and reform policies have had regressive socioeconomic impacts in an objective, concrete sense. This aspect of elitism has been the most debated and the least researched. What scanty evidence exists indicates that the net distributional impacts of many environmental policies have been inegalitarian or regressive. While Morrison and Dunlap (1980) recognize some tendency toward regressive impacts of environmental policies, their review of the literature indicates that the net distributional impacts of environmental policies vary widely from program to program. More importantly, they observe that environmental proposals do not <u>inherently</u> require regressive redistributions of economic resources, and that greater attention to distributional impacts can prevent environmentalist strategies which harm labor and the poor.

An Overview of Research on Labor and the Environment

In this section of our introductory essay, we provide a summary of the research results that are annotated below relating to labor and the environment. Both "labor" and the "environment" will be construed quite broadly. Under the rubric of "labor" we will include research results relating to a variety of phenomena, including employment levels, the relative economic position of income strata, unemployment, and so forth. We thus will be concerned about how present and potential environmental and energy policies have affected, or will affect, working-class and poor persons, regardless of whether they are members of labor unions. Likewise, we have adopted a broad definition of environmental or energy policy, focusing on air and water pollution abatement; energy price increases; "soft," renewable energy technologies; and other issues.

Effects of Environmental and Energy Policies on Labor and the Poor

We noted earlier that one of the major controversies with regard to 1970s' environmentalism was whether environmental policy implementation resulted in a transfer of economic resources from the poor to the affluent. Such regressive impacts could occur in several ways. First, the costs of pollution control could be shifted from corporations to consumers through product price increases, thereby taking disproportionately large increments from the incomes of the less affluent to pay for increasingly expensive necessities. Second, environmental control initiatives could lead to numerous plant shutdowns and to rising unemployment among working-class families. Third, environmental policies, by exacerbating problems of stagflation and economic crisis, may cause working-class families to suffer disproportionately in terms of layoffs and unemployment.

The general conclusion from several studies on the impacts of environmental control policies and expenditures is that these programs had very little deleterious macroeconomic effect and, hence, have had only a minimal impact in terms of increasing unemployment (Carter 1974; Haveman and Smith 1978). Most researchers have indicated that the U.S. economy absorbed pollution control expenditures with few if any detrimental consequences, and that claims that the rise of environmental protection during the 1970s undermined the U.S. economy are unfounded. However, there is some disagreement on whether employment was influenced by measures to protect the environment. Some researchers estimate that environmental control expenditures led to slight employment declines (Hollenbeck 1979; Data Resources Inc. 1978), while other researchers suggest that the growth in employment in the pollution control sector of the economy led to slight overall increases in employment above those that would have occurred with no increase in environmental protection expenditures (Müller 1980).

Although questions remain, there is agreement that the burdens of environmental protection have tended to be inequitably distributed, falling most heavily on the poor (see, for example, Hollenbeck 1979). As noted earlier in this essay, however, the nature of these impacts varies greatly from program to program, and ironclad conclusions are unwarranted. Environmental control policies appear to have had two effects detrimental to the lowest strata of income earners. First, insofar as privately incurred costs of environmental protection were passed on to consumers, these price increases tended to take greater shares of the income of the poor. Second, the jobs lost because of environmental protection tended to be low-skill jobs, thereby exacerbating the employment problems of those with the fewest job skills (Hollenbeck 1979). Nevertheless, these regressive distributional impacts do not, on balance, appear to have been substantial, although the economic status of semi- and high-skilled blue-collar workers apparently was affected less than that of unskilled workers.

One of the major concerns with regard to the effect of environmental protection on labor has been that environmental control policies would force factories and businesses to close, and thus would lead to increased unemployment. Available research suggests that very few industrial establishments have closed principally as a result of environmental protection policy (see, for example, Haveman and Christianson 1979). Estimates for the first seven years of the 1970s put the number of factory closings partially or largely attributable to pollution abatement at less than one hundred, which represents a tiny fraction of industrial capacity and employment in the U.S. economy.

Perhaps the greatest impact of "environmental policy" on labor and the poor has been that of energy price increases. As a result of the Arab oil embargo of 1973-74 and the Iranian oil crisis of 1979, energy prices have increased substantially over the past decade. It should be kept in mind that energy price increases do not represent an "environmental policy" which has been consciously sought by the environmental movement, though many persons inside and outside of the movement feel that energy must become more expensive if citizens are to have the incentive to conserve. Nevertheless, energy price increases—through a combination of price decontrols, future oil import shortages or interruptions, and conscious government policies such as an energy tax—are likely to remain a central feature of energy policy in the future.

It has been clearly established that "energy rationing by price" is regressive in that energy expenditures represent a larger proportion of disposable income for the poor than for the affluent. Morrison (1978) provides a convenient summary of the vast empirical literature on this topic. It should be noted, however, that while the impacts of energy price increases are quite strongly regressive with regard to direct energy consumption, the affluent tend to be massive indirect consumers of energy—that is, consumers of energy embodied in the production, distribution, and marketing of nonenergy comodities. Therefore, the impacts of energy price increases via indirect energy consumption have been less regressive than the impacts through direct energy consumption (Berndt and Morrison 1979). Nevertheless, it is clear that low income strata have been put at a disadvantage by rising energy prices and that the environmental movement, if it is to foster enduring coalitions with labor and related organizations, must be cognizant of the problems of rationing scarce environmental resources through the price mechanism.

Like research findings on the macroeconomic effects of pollution control, the dominant thrust of available research indicates that the U.S. economy has adapted to higher energy prices in a relatively painless way (see, for example, Renshaw 1981). Most studies indicate that rising energy prices have had little adverse impact on the overall rate of economic growth and that structural problems in the economy (such as the low savings rate), rather than energy problems, account for the prolonged pattern of stagflation in the United States (Carter 1974). Moreover, several studies suggest that energy price increases have been of benefit to labor because rising energy prices have encouraged industry to employ more labor-intensive technologies than otherwise would have been the case. While these employment increases to some extent have compensated for the inegalitarian impacts of price increases on consumer goods, the adverse consumption impacts have, on balance, outweighed the employment gains associated with energy conservation (Eckstein and Heien 1978).

Substitutability among Energy, Capital, and Labor

A significant segment of macroeconomic research on the role of
energy in the economy has focused on substitutability elasticities for
energy, capital, and labor. Substitutability elasticities refer to the
degree to which one input substitutes for (displaces) or is complementary to (causes expanded utilization of) another input. If the
elasticity is positive, two inputs can be considered complements in that
greater utilization of one input tends to lead to greater use of the
other input. Conversely, if the elasticity is negative, the two inputs
can be considered substitutes; that is, greater utilization of one input
will lead to less use of the other.

Substitutability elasticities for energy, capital, and labor are
extremely important in understanding historical changes in the role of
labor as a factor of production and in understanding how future changes
in technology and energy policy might affect the demand for labor and
the welfare of American workers. A variety of retrospective studies
have moved toward a conclusion that energy and labor tend to be substitutes, as do capital and labor (see, for example, Chapman 1977; Coates
et al. 1979; Berndt and Wood 1975). In other words, greater utilization
of either capital or energy tends to lead to a reduction in labor
inputs. Employment levels and returns to labor are therefore adversely
affected when the economy becomes more capital- and energy-intensive.
The available data also suggest that energy and capital have tended to
be complements in the aggregate factor mix of the U.S. economy (Chapman
1977; Berndt and Wood 1975). That is, greater utilization of capital
tends to coincide with greater utilization of energy in productive
processes.

Although the validity of these broad relationships among the
factors of production is generally recognized, it is important to note
that the patterns of relationship are contingent upon the specific types
of energy, capital, and labor inputs that are employed. For example,
Horning (1977) has determined that electricity as an energy input
historically has been a principal substitute for labor, while fossil
fuels and labor have shown a slight tendency toward being complements.
Electricity, which is utilized heavily in the industrial sector to automate the production process, thus has tended to displace labor to a
substantial degree, while the expanded utilization of other energy
inputs has led to little or no decrease in the demand for labor.

One of the major implications of such substitutability elasticity
studies concerns the potential impacts of energy policies devoted either
to energy supply expansion or to energy conservation. While it is
frequently argued that expanded supplies of inexpensive energy are
necessary to ensure economic expansion and employment, macroeconomic
research on energy utilization suggests that policies of energy supply
expansion—for example, federal subsidization of energy production which
makes energy inputs inexpensive relative to labor and capital—will tend
to reduce the level of employment in the aggregate economy. Likewise,
the fact that energy and labor tend to be substitutes suggests that
energy conservation will result in greater labor intensity of production
activities and in expanded employment opportunities for American workers
(see, for example, Ford and Hannon 1980).

Employment Impacts of Conventional and Alternative Energy Sources

In addition to the argument that expanded energy availability will enhance economic growth and employment, a corollary argument has been that expansion of conventional energy production facilities such as electrical power plants and oil drilling operations will have a beneficial impact on employment opportunities. The energy sector is one of the most important outlets for investment capital; given estimates that a vigorous program of conventional energy supply expansion might require up to three-quarters of available investment capital within the next decade (Lovins 1977), the direct employment impacts from construction and operation of facilities for energy production are of tremendous importance to American labor.

It is generally agreed that the major conventional energy industries--petroleum and electrical utilities--are two of the most capital-intensive and labor-extensive industries in the United States. Grossman and Daneker (1979:41) report, for example, that in the mid-1970s the petroleum and electrical utility industries had a capital investment per employee of $108,000 and $105,000, respectively, compared to $19,500 in industry as a whole and $9,500 in the services sector. Thus, investments made in the traditional energy industries tend to yield very little employment; most importantly, they divert scarce investment capital from other sectors of the economy which yield five to ten times as many jobs per dollar invested as do the "hard" energy industries. Accordingly, as discussed earlier in a somewhat different fashion, it is clear that far less capital is required to conserve a barrel of oil than is required to produce a barrel of oil (or its equivalent). Moreover, investments in energy conservation industries--insulation, retrofitting, and so forth--tend to yield considerably more jobs per dollar of capital investment than do the "hard" energy industries. These data lead to a conclusion that American workers as a whole are not well served by public policies which encourage or underwrite the growth of the traditional, centralized energy industries and that these investment funds would be more effectively and productively allocated in conservation-oriented manufacturing and services.

A related line of inquiry has been to explore the employment implications of a shift toward a "soft" or renewable energy economy. The available data are in general agreement that solar, biomass, wind, and kindred energy technologies are considerably more labor-intensive than traditional "hard" energy sources. The most startling documentation of this argument has been the study by the Joint Economic Committee of Congress, Employment Impact of the Solar Transition (Rodberg 1979). The study suggested that diversion of capital investment toward soft, renewable energy sources would have a substantially beneficial impact on the level of employment and that a soft energy policy thrust would contribute greatly toward full employment in the U.S. economy. While the results of the study are generally in accord with those of comparable studies (see, for example, Grossman and Daneker 1979), two major caveats are warranted. The first is that soft or renewable energy sources vary widely in their costs and resource demands. Second, as Schnaiberg (1980) has noted, the jobs created in the soft energy sector tend to be low-skill jobs which are not highly remunerated.[8] Thus, while the future employment impacts of a transition toward a renewable energy-based economy are, on balance, positive, questions remain as to costs (and hence consumption impacts on workers) and the skill levels of potential jobs in this sector.

Change in the Composition of Output

Most attention with regard to energy policy has focused on possibilities for altering the composition of inputs—for example, whether labor inputs or renewable energy sources should be substituted for electricity in industry. However, Bruce Hannon and his colleagues at the University of Illinois have suggested that the composition of output is an equally important consideration. Hannon has emphasized two aspects of the composition of output. The first concerns the possible impacts of policies that encourage or constrain the provision of certain types of goods and services (for example, constraining automobile production or highway construction and encouraging mass transit); potential impacts with regard to both energy consumption and employment change have been analyzed (see, for example, Hannon 1977). The second focus of this line of research has been on the secondary consumption impacts of reduced expenditures due to energy conservation. Hannon (1975) has noted, for example, that energy conservation, insofar as consumer dollars are saved on energy and spent on goods and services, may not entail significant net energy savings if these consumer expenditures are made on energy-intensive goods and services. Moreover, given the tendency for energy and labor to be substitutes (Chapman 1977), the reallocation of consumer purchasing power from direct energy consumption to the consumption of equally energy-intensive goods and services would potentially dilute the otherwise favorable impact on labor expected from energy conservation policies.

Several Center for Advanced Computation studies, using an energy, employment, and dollar input—output matrix, have suggested that there are a number of changes in the composition of the economy's output that would both save energy and increase employment. Examples of economic output shifts that would yield energy savings and employment benefits are changing from plane to train transportation for intercity travel, changing from car to train transportation for intercity travel, changing from electric to gas stoves, changing from electric to gas water heating, and changing from present levels to increased levels of home insulation (Hannon 1975:49). Hannon's results thus suggest that labor has a vital interest in economic decisions that affect what goods and services are produced.

The use of input—output analysis has enabled Hannon (1975) to identify a number of dilemmas in energy conservation planning. He has noted, for example, that the most effective way to reduce energy demand in the economy is to reduce incomes. Energy conservation is less likely to be effective because it generally leads to dollar savings for consumers, and the respending of this saved income may result in more energy consumption than was the case before conservation efforts. Hannon notes, however, that there are mechanisms by which the "respending effect" could be controlled and which would not penalize low-income consumers in the process. He has advocated a tax on the BTU content of energy consumed which is tied in a progressive fashion to the consumer's wage level. This BTU tax, which would place the greatest burden on high-income consumers, would provide an incentive for the respending of dollars saved through energy conservation to be directed toward less energy-intensive (and, generally speaking, more labor-intensive) sectors of the economy.

Energy, International Economic Competition, and Labor

Most of the research results relating to energy and employment are based on macroeconomic models which view the U.S. economy as a closed system in which there is no international economic competition. It is important to note, however, that international competition plays a pervasive role in shaping industrial structure and technology. U.S. firms must deal not only with domestic competitors, but also with competitors from other advanced industrial societies and from newly industrializing Third World nations. In general, the industrial firms in the advanced industrial societies tend to compete on the basis of technological innovation and increased labor productivity, while Third World industrial firms compete on the basis of their access to cheap labor. Both forms of competition are formidable obstacles for the American labor movement. To remain competitive with firms in developed countries, U.S firms must innovate technologically to save labor and increase labor productivity; the result will tend to be capital- and energy-intensive industrial growth which is labor-extensive. The presence of cheap, abundant labor in Third World economics makes it virtually impossible for U.S. firms to compete by employing labor-intensive technologies, since U.S. industrial wages would have to fall precipitously to match the low wage levels in the underdeveloped world. International competition thus provides major potential dilemmas with regard to energy conservation and employment. On the one hand, "unilateral" environmental protection might increase the costs of U.S. manufactured goods relative to those produced in other countries, thereby leading to reduced exports, balance-of-payments deterioration, national economic stagnation, and reduced employment. On the other hand, unilateral environmental protection by the United States might encourage industrial firms to locate in countries whose governments are more "tolerant" of pollution, which would also result in declining industrial employment in the United States.

The results of studies by economists, however, suggest that environmental protection in the United States is unlikely to lead to employment declines and other serious economic problems. The additional production costs necessitated by environmental protection average only 5 percent of total production expenses, even for commodities that are most affected by environmental regulations (Walter 1975). Moreover, the impact of environmental protection on the prices of export goods may be even less because public environmental programs typically rely at least partly on subsidies to polluters to offset the costs of compliance with pollution-abatement standards (Baumol and Oates 1979:201). Second, even given increased export prices, total export revenue may increase if the percentage decline in foreign demand is less than the increase in export prices. Research by d'Arge (1971) has, in fact, indicated that environmental protection would improve the U.S. balance-of-payments situation. Third, Baumol and Oates (1979:203) have argued that to the extent that environmental protection aggravates balance-of-payments problems, adjustments through the international monetary system will tend to reduce the adverse effects on the U.S. position in international trade:

[T]he evolution of the international monetary system may itself resolve any balance-of-payments problems threatened by environmental programs. Increasing reliance on flexible exchange rates has facilitated the adjustment of imbalances in international payments and is helping to mitigate many of the financial problems that are feared of environmental policy. Adjustments in currency exhange rates can automatically take care of any potential imbalances in international payments caused by the costs of pollution control. In particular, rises in a nation's costs of production occasioned by pollution-abatement programs can be offset, in terms of their effects on financial flows and domestic employment, by a depreciation in the value of the country's currency.

Baumol and Oates indicate that the principal impacts of environmental protection on industrial employment will tend to be domestic in their origins, rather than effects related to deterioration of the international trade situation of the United States. In their view, the most significant effect on employment will be the transfer of labor from heavily polluting industries to employment elsewhere in the economy; while this transfer of labor will not likely reduce employment in an absolute sense, workers will have to absorb major "adjustment costs" such as relocation, temporary unemployment, retraining costs, and loss of seniority.

Perhaps the major concern about the international employment effects of environmental protection has been that U.S. industrial firms will begin to relocate their production facilities in countries with less stringent environmental controls. Walter (1975:129-34) argues, however, that plant location is a complex decision-making process in which many variables besides pollution-abatement costs play a significant role. Because environmental control expenses tend to account for less than 5 percent of production costs, environmental regulations are not likely to be significant in decisions about locating new plants or relocating present production facilities. Environmental regulations, in Walter's view, will have a major effect on plant location only when these regulations present a crucial bottleneck in the production process. The overriding production cost consideration in plant location decisions tends to be the cost of labor, not the cost of environmental control.

Discussion: The Future of Labor and the Environment

In the preceding section, we tried to summarize what is now a substantial body of literature on the American labor movement and America's working and poor populations with regard to environmental issues. In our summary, we made several observations. First, general public (including blue-collar) support for environmental protection remains quite high and has not been irreversibly eroded by the stagflationary conditions of the late 1970s and early 1980s. Second, the working and poor populations have much to gain from enviromental protection, especially in terms of reduction of workplace and residential pollution. Third, there are a number of public policy options which can both accomplish environmental goals and improve living standards and working conditions for workers and the poor. For example, a transition toward renewable energy sources would simultaneously increase employment and conserve scarce fossil fuels, as would a shift from private to public transportation. Fourth, what have admittedly been inegalitarian aspects of environmental and energy policy are not inherent features of environmental control and energy conservation, and these regressive aspects of environmental and energy policy can be mitigated by more creative and sensitive policy formulation.

We began this introductory essay by discussing the origins of the U.S. environmental movement and by noting the charges of "elitism" in the movement and empirical evidence for it.   Evidence of both elitism (compositional, ideological, and other) and progressivism was discussed.   Anne Jackson and Angus Wright (1981) suggest, however, that the perception of environmentalist elitism has continued to grow even though cooperation between environmentalists and labor unions has become increasingly prevalent during the last few years.   Environmentalist-labor cooperation has grown largely because of the fact that the conservative shift of American politics has made it necessary for both environmentalists and labor to seek coalitions and alliances to preserve the gains both groups have made over the past decades.

Jackson and Wright, while encouraged by the growing alliances between environmentalists and labor, emphasize the as yet limited scope of environmentalist-labor coalitions.   These alliances have largely been limited to a handful of "progressive" unions--such as the Oil, Chemical, and Atomic Workers; the United Auto Workers; the Machinists; and the American Federation of State, County, and Municipal Employees-- occasionally joining forces with the Sierra Club and a few other activist environmental groups (see also Hayden 1982).

Jackson and Wright also suggest that while a great many union leaders accept the notion that environmentalists and labor can be consistent and mutually beneficial coalition partners, neither environmentalists nor labor leaders have worked hard enough to build support among labor rank and file.   "Too often, in the eyes of rank and file union members, environmentalists are spoiled rich kids who only surface when they want to block a project that could mean jobs for workers" (Jackson and Wright 1981:29).   They see an even more dismal situation with respect to minorities: "for example, the National Association for the Advancement of Colored People, concerned about unemployment, recently came out in favor of nuclear energy, despite convincing evidence that soft energy would produce more jobs than nuclear power" (Jackson and Wright 1981:30).

Jackson and Wright argue that, since the advent of the Reagan administration, the environmental movement increasingly faces "a critical juncture, a strategic and philosophical moment of choice" (1981:30).   Given the fact that continuation of the environmental movement's independent course of the 1970s would mean "the almost certain loss of environmental gains made during the last decade," the movement must inevitably join forces with either capital or labor in order to create a long-term base of support.   A coalition with capital would make possible some limited environmental gains such as resource conservation through higher energy and raw materials prices, which would ultimately lead to a squeezing of the consumption levels of workers and the poor.   Jackson and Wright reason, however, that a coalition with labor promises greater long-term gains for environmentalists, although this coalition would necessitate a major shift in strategy toward grass-roots community organizing and political education.   To mobilize the major political strength of labor--the force of numbers--attention must be shifted from lobbying and the courts toward door-to-door canvassing in election campaigns; public meetings with PTAs, garden clubs, union locals, taxpayer groups, and citizens' groups; and other techniques of building grass-roots support.

Jackson and Wright also see some potential in building a successor to the Citizens' Party which mobilized in 1980 behind Barry Commoner as its presidential standard-bearer.   They argue, however, that the

Citizens' party, with its professional middle-class image, is unlikely to achieve long-term success in American politics. Instead, they suggest that a union- or labor-based party which is cognizant of the common origins of environmental degradation and economic inequality and also cognizant of the common goals of environmentalists and workers would have the greatest probability of success.

The premise behind the compilation of this annotated bibliography was that a possible coalition between environmentalists and labor must be rooted in an underlying agenda and program which would effectively unite middle-class (and not-so-middle-class) environmentalists, workers, minorities, and the poor.[9] Such a program, if it is to be successful, probably cannot rely on either middle-class idealism, on socialist purity, or on the substantially discredited post-World War II programs of the Democratic party. We feel that the knowledge summarized below contains the building blocks for a program which is based on environmental protection <u>with</u> economic prosperity and economic security, and which is sensitive to public concerns over levels of taxation, inflation, and government regulation. We hope that this publication will serve as a clarion call for environmentalists and labor to work toward common interests, especially given the pivotal economic and political conjuncture now presenting itself in the early 1980s.

Notes

1. The expressions "labor" and the "labor movement" are not restricted to the leadership and membership of labor unions. It is widely recognized that labor unions represent a relatively small and decreasing proportion of U.S. workers. Also, unions tend to represent the most highly skilled and most privileged of nonmanagement employees in the private sector, while low-wage workers in the technologically routinized "competitive sector" (O'Connor 1973) typically are not members of labor unions.

2. Only citations that are not annotated below in this volume will be included in the reference list at the end of this essay.

3. Lovins (1976, 1977) coined the expressions "soft" and "hard" energy to refer to alternative "paths" of energy development. "Soft" energy depicts an energy path that is based on renewable energy sources which are deployed in a decentralized fashion, along with an aggressive policy of energy conservation which would decrease total energy consumption. "Hard" energy refers to an energy path which emphasizes continued expansion of traditional large-scale, centralized production of inanimate energy (i.e., petroleum, natural gas, electricity [especially nuclear power], and synthetic fuels from coal or oil shale) and which involves little attention to energy conservation.

4. It is also important to note that corporations also have attempted to encourage environmentalists to return to the "conservation" foci that predominated within the movement before the late 1960s. For example, the disposable packaging, metals, and beverage industries heavily financed a "Keep America Beautiful" campaign to divert environmentalists' attention away from ecological issues and toward cosmetic or aesthetic concerns.

5. It should be kept in mind, however, that energy price increases were not actively sought by environmental movement organizations. Thus, to the extent that energy price increases have adversely affected the consumption status of workers and the poor, the environmental movement was not responsible for such impacts.

6. We place this expression in inverted commas to point out that energy price increases were not caused by energy scarcity in a strict sense. Historical and political factors (e.g., the OPEC oil embargo, the 1978 Iranian crisis, and the political-economic power of the energy industries) were of principal importance in leading to increases in energy prices over the past decade (see, for example, McCaull 1976; Humphrey and Buttel 1982). Hence, while energy has been rationed by price, this rationing process did not derive from an objective need to increase these prices in order to reduce consumption. This is not to imply that inanimate energy sources, especially petroleum and natural gas, are infinite in their supplies and that conservation is not necessary. Instead, our point is that price increases during the past decade were not caused by a physical inability to produce petroleum to meet existing demand. For example, McCaull has discussed the coincidence of interest between the OPEC countries and the major oil companies with regard to the 1973-74 OPEC oil embargo. He notes that the OPEC embargo allowed the OPEC countries to "jack up oil prices to unjustifiable levels as insurance, not against an oil shortage, but against an oil glut. That glut has since developed and now floods the world market and would send prices tumbling were it not for the artificial price structure" (1975:3).

7. There is a wide agreement that the inflation component of stagflation was initially set in motion by the massive degree of deficit spending to finance the Vietnam War.

8. It should be noted, however, that many of the jobs created in the soft energy sector would be craftsman-type occupations associated with installation of insulation and solar equipment, retrofitting, sheet metal work, light construction, and so forth. These skilled trades occupations would be especially appealing to many workers who dislike the monotony of assembly line work.

9. One of the best sources of information on labor-environmentalist cooperation is Resources, a quarterly publication of the Environmental Task Force, Washington, D.C.

References*

Buttel, Frederick H., and William L. Flinn
    1978   "Social class and mass environmental beliefs: A
            reconsideration." Environment and Behavior 10:433-50.

Daly, Herman E., ed.
    1973   Toward a Steady-State Economy.   San Francisco: W. H.
            Freeman.

Daly, Herman E.
    1977   Steady-State Economics.   San Francisco: W. H. Freeman.

d'Arge, Ralph D.
    1971   "International trade, environmental quality, and
            international controls: Some empirical estimates."
            Appendix F in Managing the Environment, ed. A. V. Kneese,
            S. E. Rolfe, and J. W. Harned.   New York: Praeger
            Publishers.

Dunlap, Riley E., and Michael Patrick Allen
    1976   "Partisan differences on environmental issues: A
            Congressional roll-call analysis."   Western Political
            Quarterly 29:384-97.

Hardin, Garrett
    1974   "Lifeboat ethics: The case against helping the poor."
            Psychology Today 8:38ff.

Harry, Joseph
    1974   "Causes of contemporary environmentalism."   Humboldt
            Journal of Social Relations 2:1-7.

Hayden, Robert
    1982   "Can labor help save the Clean Air Act?" Resources 2:1,
            10.

Humphrey, Craig R., and Frederick H. Buttel
    1982   Environment, Energy, and Society.   Belmont, Cal.:
            Wadsworth.

Krieger, Martin H.
    1970   "Six propositions of the poor and pollution." Policy
            Sciences 1:311-24.

Lovins, Amory B.
    1976   "Energy strategy: The road not taken." Foreign Affairs
            (Fall):65-96.

    1977   Soft Energy Paths. Cambridge, Mass.: Ballinger.

*This list includes only those references which are not annotated below
in this volume.

McCarthy, John D., and Mayer N. Zald
    1973   The Trend of Social Movements in America.   Morristown,
            N.J.: General Learning Press.

McCaull, Julian
    1975   "Energy: The happy embargo." Environment (September):2-4.

Mitchell, Robert Cameron
    1979   "Silent spring/solid majorities." Public Opinion (August/
            September):16-55.

Morrison, Denton E.
    1973   "The environmental movement: Conflict dynamics." Journal
            of Voluntary Action Research 2:74-85.

Morrison, Denton E., and Riley E. Dunlap
    1980   "Elitism, equity, and environmentalism." Paper presented
            at the annual meeting of the American Sociological Associa-
            tion, New York, August.

Neuhaus, Richard
    1973   In Defense of People. New York: Macmillan.

O'Connor, James
    1973   The Fiscal Crisis of the State. New York: St. Martin's
            Press.

Schumacher, E. F.
    1973   Small Is Beautiful. New York: Perennial Library.

Sills, David L.
    1975   "The environmental movement and its critics." Human
            Ecology 3:1-41.

Wolfe, Alan
    1981   America's Impasse. New York: Pantheon.

Bibliography I:
Labor, Energy, and the Environment

compiled by
IRVING W. WISWALL and FREDERICK H. BUTTEL

001
Akarea, T., and T. V. Long
  1979 "Energy and employment: A time-series analysis of the causal
         relationship." <u>Resources and Energy</u> 2 (October): 151-162.

         Using dynamic time-series methods, the relationship between
         total employment and total energy consumption is analyzed. For
         the period of 1973-78, there appears to be a unidirectional
         causality, running to employment from energy, without feedback.
         This suggests that dynamic regressions of employment on energy
         consumption can be viewed as reflecting behavioral relationships
         as opposed to statistical forecasting relationships. The model
         indicates that employment substitutes for energy after a delay
         of eight months. The response decays geometrically and is
         practically complete by the end of the twelfth month. Thus,
         annual data cold not have captured the dynamic relationship
         between energy and employment. The elasticity of labor with
         respect to energy consumption is estimated to be -0.1356. Thus,
         constraints on energy supply would result in an increase in
         employment.

002
American Gas Association
  1978 <u>The Importance of Gas Energy to Labor.</u> Arlington, Virginia:
         American Gas Association.

         This monograph urges the development of domestic gas
         supplies to insure adequate supplies of energy for industry, to
         reduce dependence on foreign energy sources, and to increase the
         number of job opportunities in the U.S. A concerted effort on
         the part of labor, industry, and the government to support
         policies that encourage gas supply growth is advocated.

003
Anderson, Bernard E.
  1979 "Energy policy and black employment: A preliminary analysis."
         <u>Review of Black Political Economy</u> 9 (Spring): 214-37.

The approach of this study is to combine projections of sectoral economic activity, using Bureau of Labor Statistics information, with sectoral projections of black employment derived from trends evident in 1962 through 1974. The author then attempts to project 1980 and 1985 manpower implications of alternative national energy policies, with emphasis on their potential impacts on opportunities for the black labor force.

The analysis suggests that black workers will continue to experience a pattern of mixed gains in employment in an environment of higher energy costs. Specific sectoral projections are offered. Since this study is based on past trends projected into the future, it is not surprising that the analysis shows a continuation of historic patterns.

004
Anderson, Walt
    1977   "Jobs vs. environment: The fight nobody can win."   Cry California 12 (Summer):   23-28.

This article is a popularized account of the conflict between jobs and the environment, describing several issues in California. Anderson concludes that the issue may be a smoke screen behind which industry can continue to externalize costs while adopting a veneer of concern for working people. Nevertheless, the issue remains a powerful force in current politics.

005
Andrews, J.
    1979   "Employment, energy, and economic growth in Australia."   Social Alternatives 1 (September):   57-62.

Andrews examines the relationships between energy use, employment, and economic growth as they apply to the Australian economy. Low or zero energy growth would not solve the unemployment problem by itself; however, most new jobs have been created in the labor-intensive service sectors.

006
Bain, Don
    1978   "Wind energy: Net energy and jobs."   Rain 4 (May):   14-17.

The job and net energy impacts of implementing wind energy systems are discussed. A model is presented for estimating these impacts. The analysis suggests that even at low wind speeds, wind systems would provide a net energy payback ratio of 18:1. Significant reductions in unemployment would occur through jobs provided in manufacturing, transportation, and production. The development of wind conversion systems is faced with institutional problems such as the reluctance of utilities to incorporate wind-powered units into established grids and insufficient federal research and development funds.

007
Barth, Michael, Gregory Mills, and Chuck Seagrave
    1974 The Impact of Rising Residential Energy Prices on the Low-Income Population: An Analysis of the Home-Heating Problem and Policy

Alternatives. U.S. Department of Health, Education and Welfare. (American Statistics Index Microfiche 9308-5). Available from National Technical Information Service* as PB-245 206/8SL.

This study analyzes the effect of rapidly rising residential energy prices, specifically for home heating fuels, on the lower-income population, along with an analysis of various policy alternatives to ameliorate this impact. Home heating is discussed with respect to climate, housing characteristics, fuel type, and fuel prices. Regional variations in home heating cost increases and the problems faced by low-income households are given special attention. The analysis shows that low-income households spend an average of more than 11 percent of their income on natural gas and electricity. This compares with less than 2 percent for households with annual incomes over $16,000. Yet the poor consume only 56 percent as much electricity and 82 percent as much natural gas as the nonpoor.

008
Baumol, William J., and Wallace E. Oates
   1979 Economics, Environmental Policy, and the Quality of Life. Englewood Cliffs, New Jersey: Prentice-Hall.

Chapter 12 provides a general overview of the distribution of costs and benefits of environmental protection. The authors note that environmental programs tend to have regressive effects due mainly to the disproportionate costs borne by persons in low income brackets. Baumol and Oates acknowledge that the poor have the lowest-quality residential and work environments and hence derive substantial benefits from environmental protections. However, they indicate that the poor, while facing the highest levels of pollution, are less willing to pay for environmental protection than are the already environmentally privileged affluent income classes. The authors argue that the financing and implementation of environmental policy should take the vulnerability of the poor into account and transfer a greater share of the costs to high income earners.

Chapter 13 reviews evidence on how differential levels of environmental protection among nations affect trade and employment. Baumol and Oates argue that the "unilateral adoption" of stringent environmental standards would not adversely affect employment. They argue, for example, that unilateral environmental protection, when accompanied by depreciation in the value of currency, would be expected to improve the country's balance of payments and increase the demand for labor because of a decrease in imports.

009
Berman, M. B., M. J. Hammer, and D. P. Tihansky
   1972 The Impact of Electricity Price Increases on Income Groups: Western United States and California. R-1050-NSF/CSA. Santa Monica, Calif.: Rand.

---

* Address inquiries to National Technical Information Service, Springfield, Virginia 22161.

The effects of increased electricity prices on residential consumers of different income classes are analyzed, with the objective of estimating how a reduction in the growth rate of electricity consumption (through increased electricity prices) might be distributed among various socioeconomic groups in the residential sector.

Consumers in the $5,000-and-over category (60 percent of the population in the year studied) consumed 80 percent of the electricity demanded by the residential sector, whereas those earning less than $3,000 (17 percent of the population) consumed only 6 percent of the total electricity demanded. For Los Angeles, 1970, the ability of low-income groups to reduce consumption of electricity was found to be lower than had been predicted by previous research, which had used highly aggregated data to predict average reduction. This suggests that the ability to reduce consumption increases with income.

010
Berman, M. B., and M. J. Hammer
   1973   The Impact of Electricity Price Increases on Income Groups: A
          Case Study in Los Angeles.   R-1102-NSF/CSA.   Santa Monica,
          Calif.: Rand.

The study dealt with the likely effect of electricity price increases on various income groups in the residential sector of Los Angeles. A model of residential electricity consumption was fitted to data provided by the L.A. Department of Water and Power for the period 1970-71.

Consumption of electricity was determined to be influenced most by household income, the influence increasing exponentially with income levels. Relative to income, the burden of electricity price increases was found to fall most heavily on the lowest income groups. Low-income groups (below $5,000 per annum) constituted 31 percent of all L.A. households at the time of the study, but accounted for only 17 percent of total electricity consumption. For high-income groups (over $15,000), the respective figures were 21 percent and 41 percent. The evidence indicates that low-income groups, by contrast with high-income groups, have limited ability to reduce electricity consumption.

011
Berndt, E. R., and C. J. Morrison
   1979   "Income distribution and employment effects of rising energy
          prices."   Resources and Energy 2 (October): 131-150.

Research on the income distribution impacts of higher energy prices has largely focused on consumption effects; it has suggested that the effects are quite regressive when direct energy purchases are considered, but less regressive when indirect energy expenditures are also included. In contrast, this paper seeks to estimate the distributional effects of increased energy prices in the production sector. The analysis suggests that the aggregate labor cost share is basically unaffected, but that rising prices may have progressive impacts.

Rising energy prices appear to induce increased demand for labor and increased total income of blue-collar workers, and to induce a relative decrease in employment and total income of white-collar workers.

012
Berndt, Ernst R., and David O. Wood
    1975 "Technology, prices, and the derived demand for energy." Review of Economics and Statistics 57 (August): 259-68.

        Using data on U.S. manufacturing and industry for 1947-1971, the authors attempt to provide evidence on the possibilities for substitution between energy and nonenergy inputs. The study is supportive of the theory that energy and labor are modestly substitutable but that energy and capital are complements. From this, it can be expected that as energy use is reduced, either through conscious conservation or because of shortages and price increases, increased labor can be employed to a moderate extent to maintain production. However, it does not appear possible that increasing the level of capitalization is a viable way to maintain production.
        The study, unfortunately, used data from a historical period when energy was abundant and prices were low and falling. Under such circumstances, there would appear to be little incentive to develop highly labor-intensive production methods. In fact, the opposite would appear to be the case now. Similarly, little incentive existed in this period to develop technology that conserves energy in the production process. The generalizability of the study to the current situation characterized by shortages of energy with accompanying high and rising prices may be limited. Cross-national evidence from countries with high and rising energy costs that could provide corroborative or conflicting evidence was not examined.

013
Berndt, Ernst R., and David O. Wood
    1977 Consistent Projections of Energy Demand and Aggregate Economic Growth: A Review of Issues and Empirical Studies. MIT-EL-77-024wp. Cambridge, Mass.: MIT Energy Laboratory.

        Recent advances in modeling and estimation procedures permit examinations of the interactions among energy and nonenergy inputs, as well as of their simultaneous effects upon the composition and level of output. The purpose of this paper is to review the analytical and empirical evidence relating to alternative hypotheses including but not restricted to the total dependence (of the demand for energy on economic growth) and total independence assumptions. Section 1 examines the possible relationships between the demand for energy and aggregate economic growth. Section 2 emphasizes the critical importance of the energy-capital relationships and introduces the concept of utilized capital--a composite index of energy and capital. Section 3 reviews the empirical findings on substitution elasticities between energy and nonenergy inputs.

014
Betsey, Charles L.
   1979 "Energy policy and black employment: A response to Anderson and
         Hull." Review of Black Political Economy 9 (Spring): 256-59.

         The author presents a short critique of the methodologies
         used by B. E. Anderson and E. W. Hull in their articles appear-
         ing in the same issue.

015
Bezdek, Roger A.
   1975 "Toward manpower and energy dimensions for the federal budget."
         Journal of Environmental Systems 5: 29-38.

         Input-output analysis using 1963 coefficients is undertaken
         to estimate the manpower and energy requirements, both direct
         and indirect, of expenditures of $1 billion for 20 major federal
         programs. The model also permits projections of five types of
         energy needs and labor requirements disaggregated into 185
         occupational groupings.
         Caution must be exercised in interpreting specific outcomes
         of the analysis since the data are out of date and the model has
         a theoretical, but untested, assumption that relationships are
         linear. With these caveats in mind, it would appear that man-
         power and energy requirements of alternative federal projects
         vary widely.   While no clear pattern emerges, several high
         energy/low employment projects exist, as do many low energy/high
         employment options.
         From these data, Bezdek concludes that the distribution of
         federal expenditures can have a significant impact on the level
         and structure of the nation's manpower and energy needs. Energy
         and labor intensiveness of federal projects can, and should be,
         incorporated into the decision-making process.

016
Bezdek, Roger A., and Bruce Hannon
   1974 "Energy, manpower, and the Highway Trust Fund." Science 185
         (August 23): 669-675.

         Using an energy and manpower computer simulator developed
         by the Center for Advanced Computation, the authors examine the
         net impacts on energy consumption and on the labor force
         resulting from reallocating funds earmarked for 1975 highway
         construction ($5 billion). The alternatives examined are (1)
         railroads and mass transit, (2) educational facilities construc-
         tion, (3) water and waste treatment facilities construction, (4)
         law enforcement programs, (5) national health programs, and (6)
         tax relief programs.
         The authors' calculations show that manpower requirements
         of all alternatives are greater than for highway construction,
         while energy requirements are less for all except law enforce-
         ment. Railroads and mass transit (which the authors claim are
         direct substitutes for highways) would use 61.6 percent less
         energy and 3.2 percent more labor than highways. National
         health programs would use 64 percent less energy and 64 percent
         more labor than would highways.

The authors claim that energy conservation is not a serious goal of the federal government because alternatives exist but are not undertaken.   They ignore, however, the ambivalence government officials may experience resulting from conflicting goals, the difficulty of mobilizing a $5 billion dollar change in appropriations and programs in one year and the problems associated with abandoning the huge capital investment represented by the highway system.   However, they do make a clear case for the existence of low energy use/high labor alternatives to the present capital-intensive, labor-extensive approach.

017
Binder, Gordon
   1976   "Clean air, water versus jobs: A muddy issue."   Conservation Foundation Letter (July):   1-8.

   This article argues that few plants close solely because of environmental expenditures.   The threat of job loss is more frequently a scare tactic employed by industry in order to avoid paying legitimate expenses.   The article concludes that environmental protection expenses should be viewed as one of many normal costs of doing business.

018
Bjork, Lars E.
   1975   "Work organization and the improvement of the work environment." Ambio 4 (1):   55-9.

   Bjork argues that there are two principles of organizing the workplace (i.e., the division of labor):   The machine principle is that man is a substitutable part of an organization with one or a few functions.   The life principle regards man as a living being with his own intentions, ready to take on a variety of functions.   The machine principle makes people interested only in pay, not the work environment.   Programs for change are often resisted.   This lack of participation leads management to increased control over workers, leading to even less participation.   The life principle, on the other hand, leads to increased self-respect and the quest for job satisfaction.   This quest for job satisfaction also makes the work organization more adaptable to change.
   The life principle, stressing the rights of individuals, re-creates conditions whereby workers are more prone to initiate improvements and also more apt to actively support these improvements once they are initiated.   Bjork argues that shop floor autonomy, industrial democracy, and "team construction" (vs. the assembly line) point in the direction of the life principle.

019
Bloom, Martin
   1975 A Study of the Effects of Rising Energy Prices on the Low and Moderate Income Elderly.   PB-244 200/2SL.   Springfield, Va.: National Technical Information Service.

This report investigates the effects of energy costs on the income and expenditures of the low- and moderate-income elderly. Nationally, the elderly poor consume less energy than any other age-income group. Energy expenditures increase gradually as income level rises for all ages combined, but for the age-group 65 and over, the increase is dramatic from the lower-middle income level to the upper-middle income level. There were smaller differences in expenditures across income levels for natural gas relative to electricity and gasoline. Of the three energy sources, the energy gap was greatest for gasoline.

For all U.S. regions, lower-income elderly couples spent a disproportionate amount of their budget on fuel and utilities, compared to similar intermediate- or higher-budget households. The reverse was found regarding expenditures on transportation. Elderly households spent a much higher portion of their budget for energy in colder than in warmer regions. Energy price inflation hit hardest in the New England and Middle Atlantic States, and least in the South and Southwest. Overall, the rapid rise in energy prices imposed a severe economic strain on the elderly.

020
Bosworth, Barry
1976 "The issue of capital shortages." In U.S. Economic Growth from 1976 to 1986: Prospects, Problems, and Patterns, Vol. 3: Capital, pp. 1-15. Congressional Joint Economic Committee, 94th Congress, 2nd session. Washington, D.C.: U.S. Government Printing Office.

Bosworth's study supports the argument that the slow rate of growth in the capital goods sector is not significantly due to environmental policy. Imbalances and inefficiencies resulting from persistent inflation and recession are the root of the problem. Bosworth argues against relaxing environmental standards as an approach to economic problems. He suggests that capital shortages should be addressed with a fiscal policy aimed at consistent, noninflationary expansion of aggregate demand, resulting in a stable investment decision-making environment.

021
Brennan, Peter
1976 "Jobs and energy: Two sides of the same coin." Speech delivered at the 1976 Pennsylvania Electric Association Annual Meeting, Philadelphia, September 23; published in Vital Speeches of the Day 43 (December 1): 115-117.

Brennan uses statements like "energy is only the capacity to do work, but it is labor that puts the energy to work" to argue that increasing the level of employment necessitates the use of increased energy inputs. He claims that since management and capital are means of bringing energy and labor together, labor, management, and capital all share economic difficulties caused by energy problems. To solve them, Brennan has been involved in organizing the New York State Committee for Jobs and Energy Independence, basically a coalition between labor and management. Actions taken by the group include lobbying to remove environmental regulations that impede domestic energy development.

022
Briggs, Jean A.
  1977   "The price of environmentalism--The backlash begins."   Forbes (June 15):   36-40.

     Briggs documents the beginning of the decline of the environmental movement, predicting that while environmental concerns will continue to be of some importance, the public will increasingly seek a balance between ecological and economic goals.
     According to Briggs, environmental groups demand complete loyalty from political candidates they support.   This ignores the political reality that requires balancing of conflicting societal goals.   The public has become impatient with the absolute demands of environmental groups, contributing to their waning political power.
     The author notes that the 1946 forerunner of the Full Employment Act set a policy goal of 100 percent employment. Like zero unemployment, zero environmental damage is a direction society can proceed toward, but neither should be viewed as a categorical imperative.   The environmental movement is predicted to lose the extremist, absolutist tinge and to form compromises between economic and ecological goals.
     While Briggs is undoubtedly correct in the assessment that a large part of the environmental movement was extremist in the mid-1970s, the segments of the movement that were moderate are ignored.   Similarly, Briggs falls into the trap of assuming that employment and environmental goals are necessarily exclusive and ignores the important work suggesting that environmental policy can be neutral or beneficial in terms of employment.

023
Brooks, David B.
  1979   "Conserving energy creates jobs."   Perception 2 (July/August): 31-33.

     An analysis of the impact of low energy growth on the Canadian economy is briefly reviewed.   The study refutes common charges that energy conservation will result in lower employment and increased inflation.   Long-term options for energy conservation in Canada are reviewed.

024
Brooks, L. G.
  1979   "Nuclear energy = More jobs"   Energy Manager 2 (July):   21-23.

     Brooks is critical of an earlier antinuclear article (Elliott, 1979) that supports alternative energy sources because these alternative sources are more labor-intensive than nuclear power.   Brooks points out that capital-intensive production systems may employ less labor per unit of output, but they also produce more output and income per worker than do labor-intensive systems. He also points out that what may be true for a part may not be true for the whole. While an industry may provide more jobs by shifting to less capital-intensive production methods, if the whole society shifted to less capital-intensive methods, the service and public administration sectors might suffer a shortage of funds, leading to a net decrease in employment. He points out that several examples exist in economics where microeconomic elements do not add up to the expected macroeconomic outcome.

025
Bullard, Clark W. III
    1977 "Energy and employment impacts of policy alternatives." Paper presented at the annual meeting of the American Association for the Advancement of Science, Denver, Colo., February 20.

         This paper summarizes results of energy and employment impact analysis performed by the University of Illinois Energy Research Group during the period 1970-77. See the numerous articles coauthored by Bruce Hannon for more detailed accounts of specific projects.

026
Bullard, Clark W., III, and Robert A. Herendeen
    1975 "Energy impact of consumption decisions." Institute of Electrical and Electronics Engineers Proceedings 63 (3): 484-493.

         This report represents an attempt to determine the energy cost of goods and services, and was largely based upon a 360-sector input-output analysis of the U.S. economic system. The model is applied to illustrative problems, including (1) total energy cost of an automobile and an electric mixer, (2) energy impact of urban bus and auto transportation, (3) total energy impact of a family's expenditures, (4) energy and labor impacts of government spending, (5) industrial energy dependence, (6) national import-export energy balance, and (7) an energy-conservation tax. Secondary data are used and are taken from various statistical sources for the year 1963. A set of tables summarizes results of the analyses of the seven problems listed above.
         Regarding the energy impact of a family's expenditures, for the lowest-income group energy purchases account for two-thirds of the total purchases, while for the highest income group, the fraction drops to one-third. Estimates of the impact of direct energy use only might therefore be misleading.

027
Carter, Anne P.
    1974 "Energy, environment, and economic growth." Bell Journal of Economics and Management Science 5 (Autumn): 578-92.

         This study employs a closed dynamic input-output model to appraise the effects of specific pollution-abatement and new energy technologies on the rate of economic growth in the U.S. economy over the next 10-15 years. Three sets of structural

changes are examined: (1) expected changes in technology for
electricity generation, transmission, and distribution, (2)
reliance on coal gasification and fossil fuel consumption, and
(3) pollution abatement to meet stricter standards for six types
of air, water, and solid waste pollution.

When examined separately, none has striking effects on the
economy, although all do moderately depress growth rates.   The
computations show that the effects of increasing pollution-
abatement and energy costs can be offset by modest decreases in
growth of energy consumption and/or increases in the rate of
consumer savings.   In fact, an important outcome of the study is
that growth potential is more sensitive to changes in the
savings rate than to increasing costs in the energy sector.

Postwar innovations in industry were motivated by the fact
that the economy was growing at about 3.5 percent per year,
while the labor force was growing by only 1 percent annually.
Labor-saving devices had to make up the difference.   While this
trend toward labor-saving technology continues, there is an
added need to innovate because of shortages in energy and
natural resources and because of environmental limitations.   It
might be feasible to increase efficiency of both natural
resource utilization and labor simultaneously, but to do so is
more difficult than for either problem alone.   To the extent
that natural resource scarcity slows economic growth, there will
be less pressure to increase labor productivity.   However, minor
changes in consumption could offset increased immediate and
long-term capital requirements of new technologies so as to
maintain or even increase the long-run growth rate.   Thus it is
hard to predict whether environmental and energy problems will
intensify or relax historic pressures for increased labor
productivity.

028
Chapman, Duane
    1977   Energy Conservation, Employment, and Income.   Cornell Agricul-
            tural Economics Staff Paper No. 77-6.   Ithaca, New York:
            Department of Agricultural Economics, Cornell University.

            Chapman argues that current macroeconomic concepts are
probably inappropriate in the sense that in the absence of
structural changes in the economy, rising energy prices cause
unemployment, job losses, and inflation in the short run of one
to five (and possibly more) years.   Chapman's review of substitution studies suggests that employment in productive activity
may increase to replace energy over the long run, with only
slight or no change in income levels.   Energy and labor thus are
substitutes.   Chapman also reviews input-output studies which
delineate the changes in composition of consumption associated
with decreased demand for energy-intensive goods and increased
demand for labor-intensive goods.

            It is argued that current tax policy represents a subsidy
to capital.   Since energy production is capital-intensive,
public tax policy thus leads to an artificially depressed cost
of energy.   Artificially low energy prices, coupled with employ-

ment taxation which raises the effective cost of labor, will make substitution of labor for energy more difficult. Chapman notes that current business recovery policy is based on conventional macroeconomic fiscal policy, on the decline of relative oil prices, and existent price distortions. The nature of this recovery, in his view, will make the transition toward a less energy-intensive economy more difficult.

029
Chedd, Graham
   1973 "Strike action for safety at work." New Scientist 58 (April
          26): 228-9.

        Chedd discusses the alliance of 12 environmental groups with the Oil, Chemical, and Atomic Workers Union in their strike against Shell Oil. This strike was precipitated by Shell's refusal to allow worker participation in the establishment and maintenance of health standards at the company's plants.
        This strike was the first over the principle of worker input into occupational health and safety. Shell is reported to be the only company to hold out against OCAW member input into occupational health and safety procedures. Shell fears that the union will "featherbed," e.g., split up jobs as a way to solve safety problems. The main point of contention is whether worker health and safety is solely a management prerogative or also a worker concern.
        Chedd reports that the legislative director of OCAW, Tony Mazzocchi, has long been active in related issues. As the president of the Long Island local during the 1950s, Mazzocchi was involved in the "ban the bomb" issue and in the safety of atomic industry workers. He has continued to campaign to make workers aware of dangers in the workplace and was involved in the passage of the OSHA Act in 1970. Mazzocchi blames the failure of OSHA on the fact that workers have had little input into the process and that unions have not been previously involved in this area. Chedd reports that Mazzocchi is a prime mover in forging environmentalist-labor coalitions.

030
Chemical Week
   1976 "Clean air versus jobs--A case history." Chemical Week 119
          (October 27): 50-56.

        In 1975, Dow Chemical applied to the Bay Area Air Pollution Control District for a permit to build a styrene plant. Since the plant would violate Clean Air Act requirements that new plants not significantly add to air pollution in areas not yet attaining clean air standards, the permit was denied.
        The article describes various compromises between the Act and the requirements of industry across the country, including several compromises which were offered in the Bay area and refused. Throughout the article, reference is repeatedly made to the employment-generating capability of the proposed plant. However, specification of the actual number of jobs is never made, avoiding the point that chemical plants tend to be one of

the most capital-intensive and least labor-intensive forms of
industry.

031
Cherif, M'hamed, and Guillaum Yvan
    1979 "Les energies douces et la lutte contre le chomage in Belgique:
          un exemple d'application de la methode input-output." Cahiers
          Economiques de Bruxelles 81 (in French).

              The energy crisis has stimulated research into solar energy
          applications. However, installation of solar collectors is not
          yet significant, probably because solar energy is not yet
          economically competitive with conventional sources.   Using
          input-output analysis, the authors attempt to project labor
          impacts of the transition to solar energy.   They project that
          solar production equivalent to one percent of the total elec-
          tricity used in the year 2000 would induce 390 employment
          opportunities per year in Belgium.

032
Coates, Robert, Donald Hanson, Suzanne Juenger, and Jeff Kennington
    1979 Survey of the Research into Energy-Economy Interactions. Final
          report from Planco, Inc., to Macroeconomic Analysis Division,
          Energy   Information   Administration,   Department   of   Energy,
          April. Washington, D. C.: U.S. Government Printing Office.

              This extensive report consists of two volumes. Volume 1 is
          divided into two parts.   Part 1 contains a detailed and compre-
          hensive review of recent (1960-1979) literature on and ongoing
          research into the energy-economy interface.   The review focuses
          on results of theoretical and empirical analyses of energy-
          macroeconomic interactions, the various methodologies used, and
          conceptual problems in this body of research.   The survey is
          organized around a number of energy variables such as supply and
          price of energy, world price of oil, capital requirements, etc.
          Part 2 of Volume 1 begins with an overview of energy modeling
          and is followed by detailed reviews of six representative
          energy-economy models.   Volume 2 is an annotated bibliography of
          the nearly 400 articles, books, and reports cited in Volume 1.
              The sections of the report most pertinent to energy-labor
          interactions are a brief review of studies analyzing employment
          impacts of the 1973-74 oil embargo and an extensive review of
          the studies on substitutability of energy for capital and labor.
          This latter section concludes that moderate substitutability
          between labor and energy exists, but that the evidence on
          capital and energy substitutability is mixed.

033
Commoner, Barry
    1972 "Labor's stake in the environment/The environment's stake in
          labor." In Jobs and the Environment: Three Papers. Berkeley:
          Institute of Industrial Relations, University of California.

Commoner notes that U.S. technology has been transformed since World War II so that the kind of technology (as opposed to the "quantity" of technology) is leading to increased environmental destruction. These technological changes parallel the 30 percent increase in the injury rate among manufacturing workers from 1958 to 1969. The rate of increase in job injuries tends to be highest in those industries undergoing the most rapid technological change, i.e., those industries with the highest proportions of operating expenditures allocated to research and development. If long-term chronic problems such as carcinogens were included, argues Commoner, the relationship between technological change and occupational injuries would be higher.

Commoner notes the higher legally allowable contamination permitted in workplace environments as opposed to the general environment. He sees environmental pollution and hazards in the workplace as externalities. Internalizing these externalities does involve substantial costs, but for Commoner the more important question is who has benefitted in the past from unpaid costs. He makes the argument that since environmental degradation is an externality of production, producers--rather than the worker or the general public--should bear the costs of environmental control and abatement. The author emphasizes that efforts to solve environmental problems should also include measures to ensure equity so as to decrease conflict over environmental control; otherwise, in Commoner's opinion, solutions to environmental problems will be more difficult to reach.

034
Commoner, Barry
    1973 "Work place burden." Environment (July/August): 15-33.

This article presents a case history of the 1930 discovery of worker health risk from polychlorinated bi-phenyls (PCBs) after their introduction in 1929. While these worker health risks were recognized for several decades, it was not until 1966, when PCBs were found to be a general environmental problem, that concern emerged about PCBs. Commoner writes that "a worker once exposed to PCBs might be, understandably, angered to discover from articles published in this special issue [the July/August 1973 special issue of Environment on environmental health perspectives] that the discovery of polychlorinated bi-phenyl (PCB) in fish from the Baltic Sea in 1966 attracted widespread attention among scientists," while the discovery 30 years earlier that PCBs caused chloracne (a disease among workers exposed to PCBs) did not (p. 18).

The author argues that the basis for an environmentalist-labor coalition should be forged over the environmental-health problems of the workplace, since concentrations of toxic substances in work environments are greater than those faced by the general public. However, Commoner emphasizes that workplace contaminants tend to spread to the general environment and that standards for toxic substances in the workplace are 10 to 100 times less strict than standards for the "general environment."

035
Commoner, Barry
   1977   "Energy, jobs and the cost of living."   Keynote address, Los
          Angeles County Federation of Labor Conference, "Conference on
          Energy: Its Availability and Use Is Everybody's Concern," Los
          Angeles, California, January 15.

          Commoner argues that the energy crisis is still here but
          that it is in some ways less visible than it was a few years
          ago.   Nevertheless, the energy crisis is a driving force of
          inflation, a barrier to further industrial development, and a
          primary cause of unemployment.   He does not believe it is
          necessary to sacrifice environmental quality in the quest for
          adequate energy supplies.   He reviews a report on industrial
          management   in   southern   California   which   suggests   that
          alternative fuel sources can be substituted for natural gas.
          Precautions necessary in the transition to alternative sources
          are reviewed, as well as the reasons some sources have not been
          developed.

036
Commoner, Barry
   1978   "Energy and labor: Job implications of energy development or
          shortage." Alternatives 7 (3):   4-13.

          Commoner   links   the   energy   crisis,   unemployment,   and
          environmental degradation in terms of both cause and remedy.   He
          argues that public policy for solving energy and environmental
          problems can also serve to increase employment, to benefit
          workers, and to control inflation.   Commoner identifies five
          features of a sound energy policy.   First, such a policy must
          make energy available to meet citizen needs.   Second, a sound
          energy policy should discourage the use of nonrenewable energy
          resources, since nonrenewable energy resources will exhibit
          increased prices, causing rapid inflation.   Commoner notes that
          the poor are hit hardest by inflation, since inflation leads to
          economic stagnation and growing unemployment.   Third, energy
          policy   should   recognize   capital   shortages   and   the   varying
          capital efficiency of various energy sources, and should there-
          fore emphasize energy systems that maximize capital efficiency
          and minimize demands on scarce capital resources.
          Fourth, a sound energy policy should recognize varying
          employment/capital ratios of energy systems and encourage the
          development   of   energy   systems   with   high   employment/capital
          ratios so as to increase employment.   Fifth, policy should
          encourage the deployment of energy systems which have relatively
          healthful working conditions and minimal environmental impacts.
          Commoner makes the argument that industries (including energy
          production)   which   are   energy-inefficient   also   tend   to   be
          capital-intensive and labor-extensive.   Commoner views solar
          energy as being a cornerstone of a sound energy policy in terms
          of the above criteria.

037
Council on Environmental Quality
   1980   Public Opinion on Environmental Issues: Results of a National
          Public   Opinion   Survey.   Washington,   D.C.:   U.S.   Government
          Printing Office.

This monograph provides a detailed summary of the results reported in R. Mitchell (1979).

038
Dacy, Douglas C., Robert E. Kuenne, and Paul McCoy
   1979   "Employment effects of energy conservation in the U.S.A., 1978-1985." Energy Economics 1 (October): 194-202.

Using a computer simulation of the U.S. economy that combines input-output analysis with econometric analysis, the authors estimate demand for labor under policy options designed to encourage residential insulation retrofitting and increased efficiency in industrial applications.
Assuming that the U.S. government follows moderate macro-economic policies designed to keep aggregate demand growing at historic levels, the effects of both programs are small. Retrofit insulation would result in an increase in employment from 20,000 to 200,000 jobs, while increased industrial efficiency would result in a decline of about 30,000 jobs.

039
Dacy, Douglas C., Robert E. Kuenne, and Paul McCoy
   1980   "Employment impacts of achieving automobile efficiency standards in the United States." Applied Economics 12 (September): 295-312.

A computer model developed at the University of Maryland (INFORUM) that combines input-output analysis with econometric modeling is used to project employment impacts of achieving mandated automobile efficiency standards in the U.S. The analysis suggests that employment impacts will, in part, depend on the mix of weight reduction and technological innovation employed in meeting the standards. Projecting a range of most feasible mixes and implementation timetables, the authors arrive at a range of employment impacts for the years 1980 and 1985. The general conclusion of the study is that if macroeconomic policy designed to keep the U.S. economy growing at the average rate of past years is effective, auto efficiency standards will have a moderately positive effect on employment. Some shifting of employment will occur between industries, but this effect will be modest as well. Some representative estimates of the impact (in thousands of jobs) of meeting standards, by comparison with the employment picture with no standards, are as follows:

	1980	1985
Overall impact	+60 to 30	+140
Largest negative impact (wholesale and retail trade)	−16 to −22	−174 to −183
Largest positive impact (all other 178 sectors)	+40 to +37	+221 to +248

040
Data Resources, Inc.
   1979 The Macroeconomic Impact of Federal Pollution Control Programs:
         1978 Assessment.   Report prepared for the U.S. Council on
         Environmental Quality and the U.S. Environmental Protection
         Agency.   Available from National Technical Information Service
         as PB-296 960/8SL.

              Using the Data Resources (DRI) macroeconomic model, this
         report provides estimates of several economic parameters, with
         and without environmental controls, for the 1970-1986 period.
         The study suggests that pollution control will reduce unemploy-
         ment about 0.2 percentage points below what would have otherwise
         been the case.   This is true even for a shorter period when DRI
         estimates indicate a drop in real GNP associated with environ-
         mental controls.
              Several cautions regarding the study should be mentioned.
         The study assumes that standards will be met on time, that "end
         of pipe" treatment will be the method of abatement chosen, and
         that there will be no recovery of byproducts.   These assumptions
         create certain biases in the estimates.   As with all macro-
         economic models, the data used are highly aggregated and the
         information available is of the same nature.   An updated
         version, DRI's 1981 assessment, is available from National
         Technical Information Service as PB 82-109281.

041
Denison, Edward F.
   1978 "Effects of selected changes in the institutional and human
         environment upon output per unit of input." Survey of Current
         Business 58 (January): 21-44. U.S. Dept. of Commerce, Bureau of
         Economic Analysis.

              The author examines the effects of three changes: (1)
         requirements to protect the environment from pollution, (2)
         requirements to protect the health and safety of employees, and
         (3) increased dishonesty and crime and measures to combat these.
         It is claimed that these three items have in common the fact
         that they all reduce measured output per unit of input.
              The analysis shows that the combined effect of the three
         factors was a reduction in the output per unit of input of 0.2
         percent in 1972-73 and 0.5 percent in 1974-75.   Growth in output
         per unit of input had been about 2.1 percent prior to the time
         period examined.
              It is important to note that these changes refer only to
         input and output measured by market transactions.   No attempt is
         made to estimate the benefits of the expenditures that are not
         marketable, e.g., improved health and lower worker disabilities
         from accidents.   Whether output per unit of input would still
         show a decline if these factors were taken into consideration is
         unanswered in this study.

042
Dorfman, Nancy, assisted by Arthur Snow
   1975 "Who will pay for pollution control?--The distribution by income
         of the burden of the national environmental protection program."
         National Tax Journal 28 (March): 101-115.

Dorfman and Snow estimate the personal cost of federal environmental protection programs as a percentage of family income. These estimates are provided by income levels ranging from $2,000-$4,000 per annum at the bottom and greater than $50,000 at the top. Dorfman and Snow analyze separately the three major sources of long-run environmental costs: taxes to finance government expenditures on environmental programs, the costs of controlling auto emissions, and the distribution of price increases from higher abatement costs in private industry (other than the automobile sector). They find that the distribution of these costs is regressive, falling significantly as a fraction of income as one moves from lower- to higher-income families. One of the key assumptions of the study which may affect the results is that public sector expenses for pollution control programs are financed by raising taxes (rather than cutting other expenditures) and that the additional tax revenue is generated by increasing proportionately the rates on each of the major sources of tax revenues. Thus, the authors assume that 74 percent of revenues for pollution control expenditures would derive from personal income taxes and 26 percent from corporate income taxes. Thus, the results of the study are heavily shaped by the nature of the tax system, which is only mildly progressive.

043
Dorfman, Robert
   1973 "Discussion" [of Michael K. Evans, "A forecasting model applied to pollution control costs," and C. S. Russell, "Application of microeconomic models to regional environmental quality management"]. American Economic Review 63 (May): 253-6.

Dorfman contends that studies like those which Russell and Evans have conducted, and many more like them, are misdirected. They strive for "ever increasing inclusiveness and literal realism with the ultimate goal of being able to prescribe detailed plans for environmental management and explicit predictions of consequences of different policies." He notes that such detailed models have failed in planning for marketable goods and that planning for nonmarket externalities is fraught with even greater difficulties.

Dorfman believes the difficulties will not be resolved and we will be left with inexact predictions. However, he notes that quantitative models can provide important information: they can predict the direction of a change relatively well and can determine whether that change will be substantial or negligible. Using the models to draw additional or more detailed conclusions is futile and misleading, in Dorfman's view.

044
Dorfman, Robert
   1976 Incidence of the Benefits and Costs of Environmental Programs. Discussion Paper No. 510. Cambridge: Harvard Institute of Economic Research.

Dorfman demonstrates that the relative burden of pollution control tends to fall most heavily on the poor in that poor families must pay a higher proportion of their incomes for the direct and indirect costs of environmental protection.

045
Earley, Ronald R., and Makel M. Mohtadi
   1978   "Alternative energy futures and the structure of employment in the U.S. economy." In How Energy Affects the Economy, ed. A. Bradley Askin, pp. 105-119.   Lexington, Mass.:   Lexington Books.

Earley and Mohtadi utilize a Bureau of Labor Statistics model to predict 1985 employment impacts of each of the 13 Project Independence Evaluation System (PIES) energy scenarios developed by the Federal Energy Administration. The model projected employment for 129 sectors of the economy by 470 occupations. The results reported by the authors are aggregated into employment totals for 10 major sectors and nine major occupations. The 13 PIES scenarios involved combinations of assumptions about the price of imported crude oil and about domestic energy policy, although the authors placed greatest stress on scenarios which combined energy price increases (rather than price decreases) with four "policy" conditions: (1) conservation (involving "nonprice energy demand constraints"), (2) accelerated supply conditions (involving policies to encourage domestic energy production), (3) accelerated supply and conservation conditions (combining the conservation and supply expansion assumptions of the two previous situations), and (4) "supply pessimism" conditions (involving price ceilings, unfavorable geologic discovery rates, and severe environmental constraints on domestic energy production).
The net employment impacts of an increase in the world price of crude oil from $13 per barrel to $16 per barrel are modest, involving a loss of 140,000 jobs. Most of the loss of jobs occurs in manufacturing and construction, while the services, mining, banking and finance, insurance, and real estate sectors exhibit employment increases with rising energy prices. In terms of occupational shifts, employment losses occur primarily among crafts and kindred workers, operatives, and non-farm laborers, while the only occupational category registering a significant increase in employment with rising energy prices is that of service workers.
The authors find that the employment impacts of changes in domestic policy are greater than those for price increases (although it should be recognized that a $3 per barrel increase in imported oil prices is a rather modest increment by which to generalize that energy price increases will have relatively little impact on employment). In general, the supply pessimism situation leads to massive decreases in employment of roughly 1,000,000 workers, while the conservation situation results in a very small decrease in employment. Accelerated supply and accelerated supply with conservation, according to the authors, would result in a net increase of 977,000 and 576,000 jobs, respectively.

046

Early, John F.
   1974 "Effect of the energy crisis on employment." Monthly Labor
         Review 97 (8): 8-16.

         Marginal frequency analysis of the impact on employment of
         the energy shortage from November 1973 to March 1974 is
         performed to determine the effects (direct, negative indirect,
         positive indirect, and tertiary) of the energy crisis on
         employment in the U.S. economy. The four types of effects
         relate, respectively, to the inability of establishments to
         obtain the power needed for operation, to reduction of goods
         and services output, to increased demand for alternative fuel
         sources and equipment needed for extraction, and to reductions
         in aggregate demand due to layoffs.
             The most obvious direct effect was gasoline service
         station closings and reduced hours. Other direct effects were
         well scattered, but involved an estimated 150,000 to 225,000
         jobs lost from November 1973 to March 1974. For the same
         period, indirect effects entailed a total employment decline of
         310,000--more than half of this in the manufacture of automo-
         bile parts. Increased unemployment was heaviest among adult
         men, especially the 20-24 age group. The employment decline
         was smaller than those in major employment slowdowns and was
         also more concentrated in a few industries.

047

Eckstein, Albert J., and Dale M. Heien
   1978 A Review of Energy Models with Particular Reference to Employ-
         ment and Manpower Analysis. Washington, D.C.: Employment and
         Training Administration, U.S. Department of Labor. Available
         from National Technical Information Service as PB-279 447/7SL.

         This review was conducted to assess the usefulness of
         energy models with respect to energy-employment issues that
         have arisen as a consequence of post-1973 energy problems.
         Since perceptions of the energy problem vary widely, the first
         chapter is devoted to defining its nature. Chapter 2 reviews
         the main employment-energy issues. General conclusions from
         the model review are presented in Chapter 3, while Chapter 4
         discusses issues that warrant further research. The individual
         models are reviewed in an appendix.
             Some of the main conclusions of the study are that (1)
         higher energy prices will moderately retard economic growth,
         and (2) the impact of energy prices on employment will be main-
         ly through two channels--reduced output and slower GNP
         growth--which will both lead to fewer jobs. However, higher
         energy costs will lead industry to substitute labor for energy,
         thereby increasing employment. Which effect will dominate is a
         difficult issue deserving much further research. It seems
         plausible that the main employment effects will occur in the
         immediate time period as the economy adjusts to new energy
         realities. Long-run effects, while of smaller degree, will
         still be important. This energy model's usefulness in predict-
         ing these long-term trends appears limited by both the detail
         and reliability of the data bases as well as by the accuracy of

the model specifications.   The review contains an extensive
bibliography related to energy-employment modeling.

048
Eisner, Robert
    1976   "The corporate role in financing future investment needs." In
           U.S. Economic Growth from 1976 to 1986: Prospects, Problems,
           and Patterns, Vol. 3: Capital, pp. 16-32. Congressional Joint
           Economic Committee, 94th Congress, 2nd Session.   Washington,
           D.C.:   U.S. Government Printing Office.

           This study suggests that the availability of investment
    capital in the private sector is not significantly reduced by
    capital requirements for environmental control. Rather, short-
    ages of investment capital are the result of unstable macro-
    economic conditions.   The author suggests that rather than
    altering environmental goals, fiscal policy aimed at reducing
    uncertainty surrounding investment decisions should be pursued.

049
Elliott, D.
    1979   "Energy and jobs." Energy Manager 2 (April): 2-25.

           London's Centre for Alternative Industrial and Technologi-
    cal Systems has estimated the direct and indirect job impact of
    a nuclear power plant and an alternative energy-conservation
    program. Capital investment, primary energy savings, number of
    jobs created, and capital-labor ratios are calculated and shown
    to favor the nonnuclear alternative.   The article concludes
    that the long-range policy goal should be to select technolo-
    gies that balance capital and labor, resulting in conservation
    of energy and creation of employment opportunities.

050
EPA Journal (November-December, 1977)

The main theme of this issue of the 1977 EPA Journal is the relationship
of labor to the EPA.   Summarized below are the seven articles dealing
with this relationship, listed in the order in which they appeared in
the journal.

        Costle, Douglas M.
           1977   "Labor and EPA," pp. 2-3.

           This EPA administrator believes that in the next 10 years
    the agency will be primarily involved in the control of toxic
    substances.   Since these substances damage not only the envi-
    ronment, but workers' health as well, the author believes that
    EPA should be viewed as working for laborers, not against them
    as sometimes portrayed.   EPA and the Department of Labor are
    studying a proposal to assist workers dislocated because of EPA
    regulations.   The author believes this proposal should be
    extended to situations where environmental controls played a
    significant, but not necessarily determining, role in plant
    closures.   Also noted is the large number of jobs created
    directly and indirectly by EPA requirements.

The author pledges a number of actions designed to make the EPA more responsive to labor, such as more labor input into EPA policy, increasing the availability of information to workers, etc.

1977   "Cleaning up produces jobs," pp. 4-5.

This article describes numerous EPA programs' impact on jobs.

Donahue, Thomas R.
1977   "Environmental and economic justice," pp. 6-7, 26.

The author, executive assistant to then AFL-CIO president George Meany, details the history of involvement by the AFL-CIO and its predecessors in environmental issues dating back to 1908 and continuing to the present. The author presents organized labor as firmly supportive of environmentalism's spirit though not always of particular issues. Donahue expresses present AFL-CIO policy that: (1) protection of the environment can and must be reconciled with employment and energy requirements that are necessary for economic progress, and (2) the AFL-CIO is firmly opposed to policy that would move the economy to a "disastrous no-growth posture." The movement toward a clean environment and full employment will result in inevitable trade-offs, but neither side should take too rigid a stance, in the author's view.

1977   "Workers' environmental protection" (interview with Dr. Irving Selikoff, Director of Environmental Science Laboratory), pp. 8-11.

This interview provides a cursory discussion of toxic substances in the workplace and efforts by EPA, OSHA, and HEW to deal with these substances.

1977   "Environmental jobs for minorities," p. 11.

As documented in other articles in this issue, EPA programs have created thousands of jobs. However, as in other employment areas, minorities have thus far failed to get their fair share of these jobs. To help remedy this problem, EPA has awarded a grant to the National Urban League to conduct a survey of environmental job recruitment and training programs in the Northeast. With the information this survey will provide, training and recruitment aimed at minority groups can be designed.

Corrado, Frank
1977   "A union's fight for clean air," p. 12.

A feature article on a local union official of United Steel Workers who is also a dedicated environmentalist, actively working to make the steel industry less environmentally destructive and to provide a healthier work environment.

1977 "Urban workshops," p. 13.

This article describes a series of regional workshops held on the general topic of environment, jobs, and the economy. In general, the participants seem to argue that while there will sometimes be conflicts between environmentalists and labor, the two will also find grounds for mutual support.

051
Evans, Michael K.
1973 "A forecasting model applied to pollution control costs." American Economic Review 63 (May): 244-52.

Evans utilizes the Chase Econometrics model to estimate the effects which cost increases due to pollution abatement in specified industries would have on the economy. He points out that the effects are estimates because the expenditures necessary to meet regulations are estimates as well. He suggests, given this caveat, that on an aggregate level, the effects are likely to be basically absorbed by the economy without significant perturbations. While certain industries will be more affected than others, changes do not appear severe enough to cause major shifts in consumer or investment decisions.
While Evans does not explicitly address the labor-pollution control connection, the study does provide support to those who would allay labor fears that pollution control will cause major changes in industry, resulting in either reductions in investment and jobs or in dislocations within the labor market.

052
Everett, Michael
1977 "Benefit-cost analysis for labor-intensive transportation systems." Transportation 6 (March): 57-70.

Traditional benefit-cost analysis is biased against labor-intensive transportation systems like bikeways and pedestrianways, in part because they focus on narrow private transportation savings. Environmental damages (i.e., air and noise pollution) and damages to community are usually not netted out of the traditional system costs, and benefits associated with increased recreational and exercise opportunities are not included in bikeway and related systems. Everett presents an expanded benefit-cost framework and applies it to two case studies. The analysis yields benefit-cost ratios higher than found in most public projects.

053
Felix, Fremont
1976 "Electricity spurs jobs, productivity." Electrical World 186 (December 15): 58-62.

This article projects the relationship between energy use, jobs, and GNP through the year 2000. The analysis assumes that the rate of growth in energy consumed per job will rise through the end of the century at the same rate as in the 1950-1975 period -- i.e., 1.8 percent per year. GNP produced per job is assumed to grow at 3.1 percent annually for the same period, a somewhat higher rate than the rate of growth in energy per job. This can be achieved only by increasing efficiency in energy use. However, the total energy used continues to rise in this scenario to about 108 quads in the year 1985 and to 156.5 quads in 2000.

Since electrical energy exhibits greater economic efficiency in end use than do other energy sources, the bulk of additional energy must be provided in this form for higher efficiency goals to be realized. Fuel sources for this increasing energy use are assumed to be from coal, nuclear, and some additional "new sources," with a decrease in use of oil and natural gas.

This study is superficial in that it assumes all energy use is used in production while, in fact, a large proportion is utilized in residential and transportation settings. The increased efficiency of 17.5 percent in energy use that the study sees as necessary may be easily exceeded in nonindustrial applications, and electricity demand may decline (as it has, in fact, in recent years). Also, the study does not take into account shifts away from energy-intensive production in response to increasing prices and shortages of energy, nor does it consider the possible substitution of labor or capital for energy. In sum, this article can be viewed as an industry promotion for continued expansion of energy use, thinly overlain by concern for energy conservation and job protection.

054
Fiester, K.
    1977 "Energy war is generating jobs." Worklife 2 (October): 14-20.

A labor-intensive society can create many new jobs, but this fact has been overshadowed by fears of shortages and discomforts that may result from the transition to a less energy-intensive production system. Many of the jobs created by energy conservation programs have been for low-skill workers--for example, insulation and weatherization workers. Long-term employment opportunities will be for more highly skilled workers in fields such as solar, wind, and coal production. Projected estimates of the employment and energy implications of existing and proposed projects are offered to illustrate that conservation and employment goals are complementary.

055
Folk, Hugh, and Bruce Hannon
    1974 "An energy, pollution, and employment policy model." In
          Energy: Demand, Conservation, and Institutional Problems, ed.
          Michael Macrakis, pp. 159-173. Cambridge, Mass.: MIT Press.

A national linear input-output policy model, developed at the Center for Advanced Computation at the University of Illinois at Urbana-Champaign, is described. The purpose of this computer simulation model is to evaluate the effects of a

shift in expenditures at either an individual or national level
on employment, energy use, and pollution. The limitations of
the model are its linearity and the out-of-date (1963)
input-output coefficients used.

The model predicts the ratios of direct and indirect
labor, energy, and capital requirements and pollution outputs
(by type) to dollars of final demand in any of 360 industrial
sectors. At the time of the writing, the model was not complete, but some preliminary simulations had been performed.
The model predicts, for example, that a 10 percent increase in
delivery to final demand for automobiles results in a 34×10^{12}
BTU increase in direct and indirect energy use and a loss of
104,000 jobs. By contrast, a 10 percent increase in postal
services would decrease energy consumption by 4×10^{12} BTU and
create 36,000 jobs. The pollution effects of these changes
were not reported, perhaps because that part of the model was
not yet complete.

This model, if proven accurate, helps support the argument that increased production in many heavy industries
actually decreases total employment rather than increasing it.
Once the pollution prediction component of the model is implemented, the labor-environment-energy connection will be more
complete.

056
Ford, Charlotte, and Bruce M. Hannon
  1980 "Labor and net energy effects of retrofitting ceiling insula-
         tion in single-family homes." Energy Systems and Policy 4:
         217-237.

Using input-output analysis, the authors perform a detailed analysis of the energy and labor implications of retrofitting single-family houses in the U.S., assuming a seven-year
implementation timetable. Calculations are based on homes
uninsulated as of 1975. Effects on energy and labor demand in
the insulation industry and in the wider economy are examined,
both during and after the retrofit time period.

Transitional effects during and shortly after the seven-
year retrofit process result in a slight net decrease in
employment as capital is withdrawn from other sectors to invest
in home insulation, and from a direct loss of jobs associated
with reduced demand for energy. Dollar savings from reduced
energy requirements are assumed to be used to rebuild savings
or to repay loans, resulting in an increased availability of
capital in other sectors. This results in an increase in
employment in other sectors, offsetting the initial decline in
employment. After transitional effects have passed, the
authors estimate that a net increase of 28,400 jobs and a
reduction of 0.36 quads of energy per year would result from a
national insulation retrofit program.

057
Freeman, A. Myrick III
  1972 "Distribution of environmental quality." In Environmental
         Quality Analysis, ed. A. V. Kneese and B. T. Bower. Baltimore:
         Johns Hopkins University Press.

Freeman reports air pollution exposure indexes by income size classes for census tracts in Kansas City, St. Louis, and Washington, D.C. The results show a consistent tendency for levels of exposure to suspended particulates and sulfation in urban census tracts to be inversely related to the level of average family income in these tracts.

058
Freeman, A. Myrick III
    1974 Evaluation of Adjustment Assistance Programs with Application for Pollution Control. Report prepared for the U.S. Environmental Protection Agency. Available from National Technical Information Service as PB-239 423/7SL.

The federal government has considered a program to assist firms, individuals, and communities adversely affected by plant closings due to environmental regulations. The author of this report evaluates the proposed assistance program in light of expected economic impacts, including regional multiplier effects. The study concludes that the problem is not sufficiently serious to warrant a special program, especially given existing unemployment insurance, employment, and training programs.

059
Freeman, A. Myrick III
    1977 "The incidence of the cost of controlling automotive air pollution." In The Distribution of Economic Well-Being, ed. F. T. Juster. Cambridge: Ballinger.

Freeman reports data indicating that the control of automobile emissions tends to result in a strongly regressive pattern in which low-income households forfeit a larger proportion of their incomes for air pollution expenditures than do affluent households.

060
Fried, Edward R., and Charles L. Schultze, eds.
    1975 Higher Oil Prices and the World Economy. Washington, D.C.: Brookings Institution.

This book is the result of a series of papers commissioned by Brookings Institution to examine the effects of increased oil prices that began with the October 1973 oil embargo. The effects on labor are not examined at any length, however. A major conclusion of the study is that, especially in the industrialized nations, increased oil prices have not had serious or long-range effects on the world economy. In adjusting to the shock, however, the choice has been between maintenance of output and employment vs. price stability (i.e., controlling inflation). Since stability of prices was the more important goal, short-term declines in output and employment resulted.

061
General Accounting Office
    1976 Indian Natural Resources--Part II: Coal, Oil and Gas: Better Management Can Improve Development and Increase Indian Income

and Employment. GAO report RED-76-84. Washington, D.C.:
General Accounting Office.

This report concludes that the lack of inventories,
management plans, and expertise within the Bureau of Indian
Affairs has hindered the development of mineral resources owned
by American Indians. Failure to establish coal lease royalty
rates based on the market price of coal, inadequate monitoring
of lease provisions, and an inability to determine if
preferential hiring of resident Indians is taking place are
other reasons cited for the slow rate of energy resource
exploitation on Indian lands. The report urges a more
responsible attitude toward energy resource development on
Indian lands as a way of encouraging fuller energy development
that would also do much for improving the indigent position of
the vast majority of Indian peoples.

062
Gianessi, Leonard P., Henry M. Peskin, and Edward Wolff
    1977a "The distributional effects of the uniform air pollution policy
    in the United States." Unpublished Discussion Paper D-5.
    Washington, D.C.: Resources for the Future.

    1977b "The distributional implications of national air pollution
    damage estimates." In Distribution of Economic Well-Being,
    ed. F. T. Juster. Cambridge: Ballinger.

        These two publications detail a procedure for determining
    the benefits and costs of air quality programs by regions, as
    well as the limitations of this procedure. The authors argue
    that this procedure is warranted because the U.S. "uniform" air
    policy--a policy in which taxpayers' federal taxes are allo-
    cated to air quality expenditures regardless of the degree of
    air pollution in their residential area--necessarily means that
    net benefits are heavily dependent on where one lives. The net
    benefits are greatest for persons who live in the six or seven
    most polluted industrial SMSAs, and thus the distribution of
    net benefits by income classes is primarily due to the region
    of the country in which high and low income classes are concen-
    trated.

063
Goldfinger, N.
    1976 "Nuclear energy and jobs." Paper presented at the Conference
    on Nuclear Energy and America's Energy Needs, Washington, D.C.,
    March 18.

        The author, the research director of the AFL-CIO, examines
    the relationship of energy needs to the economy in general and
    to employment in particular. He argues that if energy needs
    are not met, industrial output will not be able to keep up with
    the needs of an expanding population, prices of goods will
    rise, unemployment will grow, and the standard of living will
    fall. He cites experts who indicate that increased energy
    output cannot be met with greater coal production, and
    concludes that expanded nuclear capacity is the only way energy

requirements can be met. He estimates that 1.5 million jobs
would result if 25 percent of U.S. energy needs were provided
by nuclear sources in the year 2000. The employment picture,
assuming abandonment of nuclear energy, is then discussed.
Goldfinger's discussion lacks analysis of indirect and induced
employment effects, and omits a wide range of possible alternative technologies.

064
Goldstein, Neil B., and Samuel H. Sage
    1978 "The Sierra Club's job package: An environmental works
    program." Nation 226 (February 11): 146-148.

        Government public works programs have failed to meet goals
of reducing unemployment, partly because of inadequate funding
and partly because projects are not well designed to increase
employment. Often these projects are environmentally damaging
as well. Led by the Sierra Club, environmental groups are pro-
posing a new set of public works designed to ameliorate environmental problems as well as provide ample employment opportunities. Programs cover such areas as railroad construction
and improvement, national park improvement, creation of urban
parks, and central city revitalization. The authors argue that
these projects would create more employment per dollar spent
than current public expenditures. Administrative support
appears to be lacking for this proposal, although the authors
claim there is broad public support for it.

065
Gorz, Andre
    1980 Ecology as Politics. Translated by P. Vigderman and J. Cloud.
    Montreal: Black Rose Press.

        This book, originally published as Ecologie et Politique
in France in 1975, explores the environmental problems of
advanced capitalist societies from an anarchist viewpoint. The
final portion of Chapter 4 discusses the potential role of
trade unions in providing rational solutions to environmental
problems. Gorz observes that in Europe, a majority of union
members are now less concerned with wages and other economic
aspects of work and are more concerned with the length of the
work week and related noneconomic features of their jobs. Gorz
agrees with this trend in worker opinion and argues that
"living better depends less and less on individual consumer
goods the worker can buy on the market, and more and more on
social investments to fight dirt, noise, inadequate housing,
crowding on public transportation, and the oppressive and
repressive nature of working life" (p. 133). The author urges
that unions attempt to widen the sphere of their activities
from their traditional "economistic" approach.
        Gorz argues that unions now represent a highly differen-
tiated class of manual, technical, and intellectual workers,
and that this differentiation is too great for unions to
exhibit unity on an immediate material basis. He argues that
such unity must be "constructed" by "systematically attacking
the roots of division from a class perspective" (p. 135). Gorz

sees two major bases for the construction of labor unity: (1)
unconditional protection of workers' physical integrity (there
is no "fair price" for employers to buy the ill health of their
workers), and (2) protection of workers' cultural integrity
(there is no such thing as an unskilled worker, only workers
who have been denied the chance to develop their skills).

Gorz recognizes that skill level differentiation is a
major problem in achieving unity among union members. However,
he feels that if white-collar/professional unions form strategy
around protection of the creativity of professional workers and
the corresponding obligations of these workers to enhance the
public interest, rather than merely working to increase the
wages of white-collar unionists, greater unity against environmental destruction can be achieved.

066
Grier, Eunice S.
   1976   Changing Patterns of Energy Consumption and Costs in U.S.
          Households.   Presented at Allied Social Science Association
          meeting, Atlantic City, September. Available from Washington
          Center for Metropolitan Studies, Washington, D.C.

          This paper is a report on the findings of two consecutive
          national surveys, conducted by the Washington Center for Metro-
          politan Studies, which examine the responses of U.S. households
          to increasing energy costs. Each was a random sample cross-
          section survey, the first (N=1,600) done in the spring of 1973
          and the second (N=3,200) during the spring of 1975.   These
          reports indicate that an energy-conservation ethic is beginning
          to take hold among U.S. households, but efforts to conserve are
          as yet meager.   Although residential energy costs have risen
          rapidly, they remain a relatively small portion of the average
          U.S. household's budget.   However, for certain categories of
          households--e.g., the poor and the elderly--this rising cost is
          a serious and growing burden.

067
Groncki, P. J., J. S. Munson, S. C. Kyle, and M. K. Reckard
   1978   Assessing the Employment Implications of Alternative Energy
          Supply, Conversion and End Use Technological Configurations:
          The Case of Firewood Versus Fuel Oil in New England in 1985.
          BNL-24058.   Springfield, Va.:   National Technical Information
          Service, February.

          This study estimates the direct labor requirements of wood
          fuel compared to fuel oil in New England. The study assumes
          that it would be possible to substitute an additional 0.1 quads
          of energy supplied by wood for oil currently being used. Also
          assumed is that 40 to 70 percent of residential and 90 percent
          of commercial firewood is purchased, with the remainder cut by
          the user. The analysis suggests that between 10,200 and 16,500
          additional jobs would become available by substituting wood for
          oil.   Effects on other industries--for example, wood stove
          manufacturing--and impacts on the economy outside of New
          England have been ignored.

068

Grossman, Richard, and Gail Daneker
   1979 Energy, Jobs and the Economy. Boston: Alyson Publications,
         Inc.

         Grossman and Daneker challenge the "industry" conception
   that increased energy production is necessary for increased
   employment and a growing economy. They note that solar and
   other alternative energy sources are labor-intensive, but
   require relatively unskilled labor. Because unskilled labor
   does not tend to be unionized, increased job opportunities in
   solar energy may threaten existing unions. Grossman and
   Daneker note that historically the U.S. has experienced substitution of energy for labor. Unions have concurred in this
   energy substitution process, opting for high wage scales tied
   to the "productivity index." The authors present data indicating that conventional energy production is capital-intensive,
   diverting capital from other, perhaps more labor-intensive,
   investments; that conservation-related industry is labor-
   intensive; that conservation is less capital-intensive than
   conventional energy production, which would help to avoid
   capital shortages; and that solar energy is labor-intensive,
   environmentally safe, and technically feasible.
         Grossman and Daneker see two major political obstacles to
   the development and deployment of solar energy: (1) the desire
   of large corporations with investments in nonrenewable energy
   to protect these investments and to retain the government
   subsidies they have enjoyed, and (2) government neglect. The
   authors argue that federal policy regarding energy and employment within DOE and DOL is uncoordinated. They also stress
   that for regulated energy production utilities, expansion of
   generation capacity and of energy production is necessary to
   increase total profits, since the rate of profit is controlled. Therefore, regulated utilities' desires to expand
   capacity and to stimulate demand to consume the available
   supply are a barrier to conservation and solar energy.

069

Hannon, Bruce
   1975 "Energy conservation and the consumer." Science 189 (July 11):
         95-102.

         The author uses input-output analysis to determine the
   energy and labor implications of consumer choices. He notes
   that a number of dilemmas emerge from attempts to lower energy
   consumption: (1) Labor appears to displace energy by decreasing the number of high wage jobs and increasing the number of
   low pay opportunities. (2) It appears that the most effective
   way to save energy is to reduce income. (3) Saving energy
   leads almost inevitably to dollar savings. Respending this
   saved income results in more energy consumption. This effect
   can result in energy use which is higher after conservation
   efforts than before. A tax based on the BTU content of energy
   consumed and tied to the consumer's wage level is proposed as a
   policy tool to ameliorate the contradictions implicit in energy
   conservation.

070
Hannon, Bruce
    1977   "Energy, labor, and the conservor society." Technology Review
            (March/April):   47-53.

        Using the input-output matrix developed at the Center for
Advanced Computation at the University of Illinois, Hannon
demonstrates that there are a number of options for change in
the composition of the economy's output which would have the
effects of both decreasing energy use and increasing employment. Examples include changing from planes to trains or from
cars to buses for intercity transit, from electric to gas
stoves, from electric to gas water heating, etc. Likewise,
other possible changes in the composition of output would lead
to more energy consumption and employment (e.g., changing from
electric commuter transit to buses), to decreased employment
and energy use (e.g., changing from present to new electricity
supplies), or to increased energy use and decreased employment
(e.g., changing from beef protein to textured soy protein).
Hannon thus suggests that public policy should not only address
the employment and energy impacts of changes in productive
inputs, but also the ways in which substitution of goods and
services in the final output mix might enable energy conservation to lead to increased employment levels.

071
Hannon, Bruce M., and Roger Bezdek
    1973   "Job impact of alternatives to Corps of Engineers projects."
            Engineering Issues 99 (October):   521-31.

        Corps of Engineers projects are often criticized as being
unnecessary, wasteful, and harmful to the environment.
Defenders of these projects refute these charges and often add
that employment increases, both nationally and in the area of
the project, are a major benefit of such projects. While
employment benefits of Corps projects are not to be denied,
comparison of employment impacts of alternative undertakings is
necessary to complete the job impact picture.
        The authors use an input-output model to estimate the
direct and indirect employment effects of Corps projects and
alternative investments. The analysis indicates that reducing
the Corps budget by 75 percent and allocating the savings of
$1.137 billion to alternative uses would create the following
employment changes:

Investment category	Employment	% Change relative to Corps projects
Corps construction, i.e., no change	73,380	--
Tax relief	81,182	+10.6
National health insurance	115,419	+57.3
Social security benefits	95,968	+30.5
Mass transit construction	78,443	+ 6.9
Waste treatment construction	95,609	+30.3

Any of the alternatives is projected to create more employment than Corps projects. Waste treatment construction, in some ways substitutable for Corps projects, would create over 30 percent more employment while at the same time avoiding criticisms which environmentalists frequently level at Corps projects.

072
Hannon, Bruce M., Carol Harrington, Robert W. Howell, and Ken Kirkpatrick
    1979 "The dollar, energy and employment costs of protein consumption." Energy Systems and Policy 3 (1979): 227-241.

The authors compare the dollar, energy, and employment requirements of producing a kilogram of utilizable protein via beef production, and via textured soybean protein and direct soybean consumption, through input-output analysis. Not surprisingly, they find that beef raised in the Corn Belt is six times as energy-intensive and six times as expensive as textured protein. Beef is nine times as energy-intensive and ten times as expensive as direct soy consumption. However, the labor requirements of producing beef protein are about 14 times higher than for textured protein. Thus a switch to soy protein would entail a certain amount of unemployment.

The authors also studied the effect of dollar savings from switching to soy consumption, assuming that the savings would be spent on average consumer goods. Since beef production is less energy-intensive than the average consumer goods, there would be no net energy savings from the switch and there could even be a small increase in energy demand. Once respending effects are taken into account, the net effects of switching from beef protein to textured soy protein would be an employment drop of about 600,000 jobs and an energy increase of about 0.12 quads. Alternatively, if respending were assumed to be on average consumer goods but excluding direct purchase of energy, employment would still decline by 560,000 jobs, but energy demand would also decline by about 0.35 quads.

073
Hannon, Bruce M., and Andrew R. Pleszkun
   1978 Energy and Labor Intensities Projected to the Year 2010.
         Washington, D.C.:   U.S. Department of Energy, Division of
         Buildings and Community Systems.   Available from National
         Technical Information Service as COO-4628-2.

         An input-output model of the energy and labor requirements
         of the economy is used to project energy and labor requirements
         under a variety of assumptions, including four energy price
         change scenarios.   The impact of the national energy plan is
         also investigated, and residential energy use is projected to
         the year 2000.   A fairly detailed description of the model is
         also presented.

074
Hannon, Bruce M., Richard G. Stein, B. Z. Segal, and Diane Serber
   1978 "Energy and labor in the construction sector."   Science 202
         (November):   837-47.

         A labor and energy input-output model of the economy
         developed by the Energy Research Group at the University of
         Illinois is employed to study the potential for energy conser-
         vation in the construction sector as well as a possible inverse
         relationship between total energy use and labor requirements in
         this sector.   Both direct and induced labor and energy require-
         ments as well as lifetime labor and energy and labor operation
         costs were considered for construction.
         Three approaches to energy conservation and the effects on
         labor requirements were analyzed: (1) the selection of less
         energy-intensive materials, (2) analysis of lifetime energy
         costs including maintenance, and (3) increasing energy effi-
         ciency in the production of building materials.
         The model reveals that total energy consumption in the
         construction sector could be reduced 20 percent by (1) using
         construction techniques and materials that result in lower
         energy needs, (2) implementing improved energy efficiency in
         materials production, and (3) using slightly more energy in
         initial construction, mainly embodied in increased insulation
         levels, to minimize total lifetime energy requirements.
         The study supports the theory that energy and labor have a
         moderate inverse relationship in the building construction
         sector.   Less energy-intensive building methods appear to use
         about five percent more labor per dollar spent than more
         energy-intensive methods.

075
Haskell, Mark A.
   1977 "Green bans: Worker control and the urban environment."
         Industrial Relations 16 (May):   205-14.

         Beginning in 1971, Australian building trade unions
         employed "green bans"--that is, boycotts of employers and
         others to protest construction projects felt to be destructive
         of environmental amenities.   Noting that American trade unions
         are conservative and often opponents of the environmental move-
         ment, the author seeks an understanding of why the Australian

labor movement was able to transcend traditional goals and combine forces with community groups to exert wider social influence.

Haskell's conclusion is that, in large part, this movement was able to emerge because of rather special circumstances. First, a large segment of the Australian labor movement is communist in orientation and internally divided. Leaders found supporting green bans to be consistent with their goals of establishing a militant reputation among the competing factions. Second, participatory democracy and involvement in wider social issues, both within unions and in the larger society, have long been a goal of the unions involved. Since bans were imposed only with wide community support, green bans were a convenient vehicle for expression of this ideology. Third, unemployment at the time was very low. Finally, the participating unions had large turnover rates, precluding the organization of internal opposition to green bans.

Despite the rather specialized circumstances of the emergence of green bans, ideological considerations did play a rather central role. Haskell is thus cautiously optimistic that similar action has potential in other countries.

076
Haveman, Robert H.
   1977   Employment and Environmental Policy--A State-of-the-Art Review and Suggested Approach. Report prepared for the Organization for Economic Cooperation and Development, Environment and Industry Division, Environment Directorate (Paris).

The interdependent channels through which environmental policy affects the level and structure of employment are explored. The author points out that in a large and complex economy, the net employment impacts are difficult to estimate. A research approach designed to provide estimates of both direct and indirect employment impacts is presented. It is suggested that OECD might profitably develop a multicountry research effort designed around a shared methodology. An appendix summarizes several studies in the U.S. and abroad that investigate employment-environmental policy interactions.

077
Haveman, Robert H., and Greg Christianson
   1979   Jobs and the Environment. Scarsdale, N.Y.: Work in America Institute, Inc.

This publication is a major review of the literature on the interactions between environment and employment, most of which is usefully abstracted. The authors identify six channels through which environment and employment interact: (1) Environmental control may increase production costs, thereby decreasing demand and possibly decreasing employment. (2) The rise of the environmental control industry creates employment. (3) If economic growth is incompatible with a healthy economy, there is the question of whether employment would decline in a no-growth economy or whether employment could be held constant. (4) Attention to environmental problems may only be possible when there are favorable (i.e., growing) economic conditions. (5) Increasing energy prices and shortages may increase unemployment and increase the use of coal and nuclear energy, with their attendant environmental problems. (6) Workplace conditions are linked with environmental concern. Compliance with workplace contaminant standards may increase production costs, decrease demand, and result in declining employment.

The authors' literature review focuses on the first two channels. Five kinds of studies are considered, including the "bottom-up approach" of microsimulation techniques, "top-down" or macroeconomic modeling, and studies of employment impacts in other countries. In general, the first three techniques find small, but statistically significant, negative impacts of environmental regulation on employment—in the area of 0.2 percent. The fourth type of study identified by the authors is analysis of the effects of substituting certain environmental spending for other spending. By decreasing spending on labor-extensive programs and increasing spending on labor-intensive environmental projects, increased employment is considered to be plausible.

The final type of study is that of sectoral impacts of environmental control. The pollution abatement industry, construction, and the public sector are found to experience increases in employment as a result of increased spending on environmental control. From 1971 to 1976, 98 U.S. plant closings, with a loss of 19,580 jobs, have been attributed, at least in part, to increased costs of environmental control. Many of these plants were economically marginal, however. Finally, the authors project that GNP will decline with increased environmental control spending. Nevertheless, the effect of pollution control spending on GNP growth in the past decade has been quite small; the authors attribute the slow rate of capital formation in recent years largely to imbalances and inefficiencies caused by inflation and recession.

078
Haveman, Robert H., and Julius Margolis, eds.
   1970 Public Expenditures and Policy Analysis. Chicago: Markham Publishing Company.

   All of the articles appearing in this collection of papers appeared in U.S. Congress, Joint Economic Committee, The Analysis and Evaluation of Public Expenditures: The PPB System. Washington, D.C.: U.S. Government Printing Office.

079
Haveman, Robert H., and V. Kerry Smith
   1978 "Investment, inflation, unemployment, and the environment." In Current Issues in U.S. Environmental Policy, ed. Paul Portney et al. Baltimore: Johns Hopkins University Press.

   The authors assert that "trade-offs are necessary between environmental objectives and a broad range of other social objectives—full employment, price stability, energy conservation and economic growth to name but a few." In order to assist in making these trade-offs, various models that predict the aggregate economic impact of environmental policies have been developed. The article describes, in general terms, various approaches that have been taken to explore this problem. They compare a few specific models that are representative of broad categories of approaches, and an attempt is made to assess their reliability. Estimates from the studies are used to evaluate a number of assertions regarding the impact of environmental policy in the U.S. economy.

In general, the models do not support the contention that increasing pollution abatement will pose substantial strain on financial markets. The employment impacts of environmental policies seem to be, at worst, minimally adverse. Several studies predict modest increases in employment. Most models show some rearrangement of the labor market, reflecting expansion and contraction of various sectors. Many models show that environmental policy can serve to stimulate employment in the short-term.

Regional vs. national impacts do appear different. Overall impacts on employment may be small, while local impacts may be great. Regional increases as well as declines may occur.

080
Heller, Walter
   1971 "Economic growth and ecology--An economist's view." Monthly Labor Review 94 (November): 14-21.

Heller's argument is that the economist's view of environmental problems contrasts with that of the ecologist's in the following ways:

(1) The ecologist sees problems in absolutes: all pollution must be stopped. In contrast, the economist is trained in marginal analysis and tends to see problems in terms of arriving at an optimal balance. In other words, pollution must be reduced to the point where the cost of reduction matches the benefit of that reduction.

(2) Ecologists believe that control of environmental problems must come about through government prohibitions and regulation, while the economist tends to rely on pricing systems as much as possible, favoring effluent charges, for example, to reduce pollution.

(3) Ecologists believe that if the externalities related to the past 25 years of growth were internalized, the increase in GNP would disappear. Heller disagrees (without providing firm evidence), saying that, indeed, the rate of growth would have been less if the health and environmental damages were deducted from the gains, but he contends that growth in the economy still would have occurred.

Heller believes that environmental deterioration must be stopped and that the funds necessary for the effort must come from continued economic growth.

081
Herendeen, Robert A.
   1974 "Affluence and energy demand." Mechanical Engineering 9(6): 18-22.

Input-output analysis was performed on 1960-61 Bureau of
Labor Statistics Consumer Expenditure Survey data for 368
sectors of the U.S. economy (aggregated to 97) in order to
evaluate direct and indirect energy needs of three income
classes.   The effect of income was measured with regard to
seven consumption categories--i.e., direct energy purchase,
food and water, housing and clothing, auto purchase and maintenance, medical and education, transportation and recreation
(besides auto), and investments.   It was found that the
indirect energy impact increased with income.   Two-thirds of
energy use was indirect for the highest income classes, compared to one-half for all consumers.   The author concludes that
a flat-rate energy tax would be less regressive than one only
on direct uses.

082
Herendeen, Robert A., and Jerry Tanaka
   1976 "Energy cost of living." Energy 1(2):   165-178.

Evaluation of energy requirements of household expenditures for all products from the 1960-61 Consumer Expenditure
Survey of the Bureau of Labor Statistics (N=13,000) is performed using input-output analysis.   Socioeconomic variables,
e.g., income, number of members, location, and age of family
head, are related to household energy requirements and expenditures.   Within error bounds, one universal curve shows the
dependence of energy impact of expenditures for households of
two through six members.   For a typical poor household,
purchases of residential energy and fuel account for about 65
percent of energy requirements.   For an affluent household,
this fraction drops to 35 percent.

083
Heritage, John
   1979 "Labor's stake in steel cleanup" (interview with Lloyd McBride,
         President of United Steel Workers of America). EPA Journal
         (March):   32-33.

McBride offers his position on a number of issues stemming
from EPA and OSHA regulations affecting the steel industry.   He
believes that these environmental and health regulations have
not been the source of recent economic problems faced by the
industry, as steel company advertising campaigns have suggested.   Steel industry problems, he believes, are the result of
larger economic problems and a failure of steel companies to
modernize production facilities in the past.   McBride also
notes some specific job protections that federal pollution laws
offer workers, especially protection from "environmental blackmail" and wage protection in cases where industry controls
pollution through production slowdowns.   He notes that state
implementation plans under the Clean Air Act offer opportunity
for union involvement in planning that should be better utilized.   Finally, he is critical of cost-benefit analysis as a
yardstick for deciding the value of environmental regulations.

084
Hill, Gladwin
    1976   "The profits of ecology: Cleaning the stables makes jobs."
           Nation 222 (April 17):   455-458.

           The author argues that environmental protection is rela-
           tively inexpensive and that the benefits outweigh the costs.
           Statistics are presented supporting the view that the
           availability of jobs can be increased through environmental
           protection programs.   The notion that environmental controls
           are responsible for unemployment, inflation, energy shortages,
           and a sagging economy are presented as scapegoating to draw
           attention from the underlying economic ills.

085
Hollenbeck, Kevin
    1979   "The employment and earnings impacts of the regulation of
           stationary source air pollution."   Journal of Environmental
           Economics and Management 6:   208-31.

           The results of Hollenbeck's research on the regulation of
           stationary source air pollution include: (1) Regulation of
           stationary source air pollution increased the consumer price
           index by 0.0002 for actual expenditures and by 0.0035 for
           projected expenditures.   (2) With few exceptions, the impact of
           pollution control on household earnings is negative.   Above the
           lowest income groups, the impact on household earnings is
           uniformly regressive.   (3) Changes in final demand and output
           are small overall.   However, wide variations in change occur.
           Industries employing the most pollution abatement equipment
           experienced small declines in demand and output while
           industrial sectors producing that equipment posted moderate
           gains.   Sectors of the economy whose demand and output are most
           closely dependent on consumer demand experienced small
           declines.   (4) Total employment declines were 73,000 to 73,440
           for actual employment and 156,810 for projected employment.
           Most of the employment impact was concentrated in low-skill and
           white-collar jobs.   Semi-skilled and high-skill workers (who
           predominate in areas closely related to production of pollution
           control equipment and in inputs to that industry) experienced
           little decline, or perhaps an increase.   (5) The regressive
           tendency of pollution control remains if household, rather than
           individual, income is considered.

086
Hollenbeck, Kevin
    1976   "The employment and earnings incidence of the regulation of air
           pollution."   Ph.D. dissertation, University of Wisconsin,
           Madison.

           The author presents a comprehensive microsimulation model
           for evaluating the economic impact of environmental policy.
           Stationary source regulations of the amendments to the Clean
           Air Act of 1970 are investigated with the model.
           The model is a partly closed, interindustry model contain-
           ing wage and price equations that estimate how wages and prices

in various sectors respond to policy changes. Once changes in
demand for goods are determined, this information is entered
into a macroeconomic production model that yields estimates of
employment demand.

Hollenbeck relies on EPA estimates of 1971-79 investment
costs and annual costs to the industrial sector that were
attributable to the legislation. The analysis suggests that
employment declined slightly (0.21 percent) as a result of the
amendments and that low-skilled workers were most seriously
affected.

087
Hoover, John L., Danilo J. Santini, Kenneth Smeltzer, and Erick
Stenehjen
    1980 "Concentrated vs. distributed energy: Employment-based com-
          munity level differences." Paper presented at the annual meet-
          ing of the American Association for the Advancement of Science,
          San Francisco, January 4.

          This paper compares the employment impacts of concentrated
          vs. distributed energy systems, focusing on one particular end
          use, space conditioning. Employment effects are considered on
          the community level, as opposed to most studies which consider
          net national effects.

          The analysis shows that direct solar heating and cooling
          of buildings have considerably higher labor requirements per
          BTU of energy produced than do conventional concentrated
          energy-producing technologies. Solar electric systems (wind,
          ocean, and thermal power, etc.) are intermediate between direct
          posted solar applications and conventional power sources in
          terms of labor input per BTU produced.

          The authors note other advantages of distributed energy
          systems. Employment is also distributed, and would thus help
          alleviate unemployment across the country. Community stability
          would be enhanced, since the boom and bust cycle of large
          energy projects would be avoided. This is especially important
          since new power projects (i.e., western coal development) are
          often located in rural areas with little capacity to absorb
          large influxes of construction workers. Finally, costs and
          benefits of distributed energy systems are more perfectly borne
          by the same people; concentrated systems often impose costs
          near the facility while the benefits are enjoyed at a distant
          demand center.

088
Hornig, Ellen C.
    1977 "The effects of energy pricing on employment in New York State
          manufacturing, 1964 to 1973." Ph.D. dissertation, Cornell
          University, Ithaca, New York.

          Several econometric studies have been conducted to inves-
          tigate the response of the demand for labor to changes in the
          price of energy and other production factors. Unlike previous
          studies, this one uses state-level (N.Y.) data, separates pro-
          duction from nonproduction labor, and differentiates between
          energy provided in the form of electricity from that used as

fossil fuel. This, the author argues, yields more information about patterns of substitution between labor and energy than the more customary use of single variables for energy and labor. A partial adjustment model is employed which does not assume instantaneous adjustment of prices as does the traditional approach of the translog form of analysis.

The major conclusion of the study is that electricity seems to have been substituted for labor between 1964 and 1973, while fossil fuels and labor have been complementary. Both elasticities are small but significant. This suggests that increasing prices of electricity should increase employment in manufacturing. Separate investigation of labor-intensive vs. capital- and energy-intensive industries indicates that this response will likely be greater in the former than in the latter. Since this labor-intensive group of industries in New York State has suffered recent competition from low-wage regions and countries, increases in electricity prices probably cannot have significant positive impacts on employment. The study concludes that the state would not be the most effective level for electricity price manipulations as a tool to stimulate employment. National electricity price increases may attain this goal, however. Evidence is presented which, nevertheless, suggests that high energy prices in New York State have not been the chief cause of manufacturing job losses.

089
Hudson, Edward A., and Dale W. Jorgenson
 1974 "U.S. energy policy and economic growth, 1975-2000." Bell Journal of Economics and Management Science 5 (Autumn): 461-514.

This paper presents a new methodology for the analysis of U.S. energy policy, consisting of an integration of econometric modeling and input-output analysis. The model is used to project U.S. economic activity and energy utilization for the period 1975-2000, initially under the assumption of no change in energy policy. The model is then used to design a tax policy that will stimulate energy conservation and reduced dependence on imported sources of energy.

The advantage of this modeling technique is that in contrast to traditional approaches, energy is not viewed in isolation from other economic phenomena. Rather, energy is seen as one of the many interacting parts of the economic system. The explicit linkage of energy development to such variables as employment, income, and consumption can thus be explored.

The major findings of the study are that with increasing energy prices, energy use does decline, but that the cost of decreasing energy use on the economy is minimal. Hudson and Jorgenson project that an increase in energy prices leading to a 7.8 percent decrease in energy use would result in only a 1 percent increase in prices generally and a 0.4 percent decline in real income. The 7.8 percent decline in energy use would be accompanied by increased labor use (0.6 percent) and increased capitalization (0.5 percent).

090
Hull, Everson W.
   1979 "U.S. energy policy and employment opportunities for the poor."
         Review of Black Political Economy 9 (Spring): 238-55.

         The recent historical record (1954-1977) regarding GNP,
         employment discrimination, and energy use is examined in this
         article.   Based on past trends, the author makes policy
         recommendations favorable to black employment.
         Hull first demonstrates that in the past, the closer the
         economy has functioned to maximum potential, the lower have
         been the rates of unemployment for blacks and the more equal
         the distribution of employment between whites and nonwhites.
         He then argues that U.S. energy policy has created economic
         inefficiencies that retard economic growth.   He also presents
         evidence on the displacement of black workers caused by energy
         shortages in the mid-1970s.
         Hull argues that decontrolling the price of energy will
         attract investment, raising the output from energy-producing
         sectors. This increased production and consumption will stimu-
         late economic activity necessary for equitable employment
         opportunities for blacks. Problems associated with high energy
         costs for low-income families should then be addressed directly
         rather than through the pricing mechanisms traditionally used.

091
International Metalworkers Federation
   1978 Energy. Geneva: International Metalworkers Federation.

         This publication was intended to provide information to
         IMF member unions regarding energy, particularly nuclear
         energy.   Its contents include:
         (1) Barry Commoner's "Energy and labor: Job implications
         of energy developments or shortages."
         (2) A chapter on "Some special characteristics of the
         nuclear safety debate" from Energy: Global   Prospects,
         1985-2000, by Carroll L. Wilson (New York: McGraw-Hill, 1977).
         (3) Results of a survey of member unions' official posi-
         tions (supplemented by research on nonrespondents) in regard to
         nuclear energy.
         In general, member unions are heavily involved in the
         nuclear industry, with 116,063 workers employed in energy
         production and 34,894 in the nuclear sector. Most unions are
         correspondingly pronuclear. However, most unions also promote
         alternative energy sources and conservation. Most unions also
         call for continued R & D to solve problems of nuclear energy,
         e.g., waste disposal and safety. Most unions link increasing
         energy use with increased economic growth, and the need to
         increase economic growth for maintenance of jobs and of the
         standard of living of workers. No union actively contests this
         thesis. The survey found that most unions call for decreased
         dependence on foreign oil.
         The Australian Amalgamated Metalworkers Union firmly
         opposes nuclear power and uranium mining for environmental and
         health reasons.   The union feels that increased reliance on
         nuclear power will cause employment to suffer and calls on the

government to promote solar and coal power instead of nuclear. Other Australian metalworker unions basically support this stand, but not as vigorously as the Australian Amalgamated Metalworkers Union.

The report includes a statement by the IMF regarding nuclear energy which includes the following points:

(1) IMF supports the concerns of all workers regarding the maintenance of employment and the right to a healthy environment.

(2) The IMF supports democratic national and international debate on energy.

(3) IMF supports the development of economic structures that allow the conservation of energy.

(4) IMF urges that R & D on nonnuclear forms of energy be increased.

(5) IMF feels that while it is not feasible to halt the production of nuclear energy, nuclear power systems should be deployed only where necessary. Safety risks of nuclear power should be decreased.

(6) IMF insists that problems of waste, reprocessing, and storage should be solved.

(7) IMF calls for public participation in nuclear plant siting.

(8) IMF insists that the safety of nuclear workers be safeguarded.

(9) IMF favors the nationalization of and public control over energy corporations of all kinds.

092
Jackson, Anne, and Angus Wright
   1981 "Nature's banner: Environmentalists have just begun to fight."
         The Progressive (October): 26-32.

Jackson and Wright, in a highly provocative article on the future of environmentalism in the 1980s, discuss the major social forces that will affect environmental protection through and beyond the Reagan administration. They begin with a paradox: The membership of environmental groups is at record levels, and public support for environmental protection remains strong, yet the Reagan administration has chosen to make the environmental movement "the target of a determined and audacious assault" (p. 26).

"With so much public backing of environmentalism and with that backing likely to intensify, why do the Reagan administration and the corporations believe they can make the nation retrace its steps to an earlier, more indifferent age?" (p. 26) Jackson and Wright argue that the answer to this question lies not in the decline of environmental concern, but rather in the rise of a virulent counterattack on environmental protection. As corporations have come to feel the impacts of environmental legislation--including both specific costs of environmental protection and larger costs of public questioning of industry's control and authority--business firms and wealthy individuals have begun to channel large sums of money into political action committees, conservative think tanks, conservative political candidates, and political advertising. A significant thrust of

"new right" activity has become an antienvironmental counterattack aimed at returning to industry the authority to determine the technologies to be used in production and treatment of
wastes.

Jackson and Wright emphasize that while industry and
conservatives have taken their message to the public at large,
environmentalists have mistakenly left the public "on the
sidelines" (p. 28). Environmentalists have tended to work
through a few front-line activists in Washington, D.C., in
state capitals, and in courtrooms. · This tendency was magnified
by the presence within the Carter administration of a number of
former environmental organization officials; environmental
organizations thus fell into a pattern of placing principal
emphasis on lobbying their former associates in the government. The authors recognize the importance of lobbying and
courtroom activities, but stress that strategies aimed at
mobilizing the public will be necessary to meet the conservative challenge.

Jackson and Wright place principal emphasis on building
coalitions with public interest groups and groups representing
workers and the disadvantaged, both in national politics and in
the door-to-door canvassing of local elections. Jackson and
Wright's comments on coalition-building focus primarily on
unions. They dispute the notion that union leaders are
generally antienvironmental, pointing out several instances in
which union leaderships have taken progressive stands on
environmental issues. The problem, in their view, is that no
one--union leaders or environmentalists--has worked hard enough
to stress to union members that proenvironmental policies tend
to have positive implications for employment and economic
prosperity. The authors approvingly note the activities of the
political action committee of Friends of the Earth in
preparation for the 1982 elections. The FOE PAC efforts and
their preliminary successes are taken as evidence that
environmentalists can effectively form coalitions with labor
and community groups to counter the political advertising of
conservative antienvironmental groups.

093
Jaroslovsky, Rich
  1976 "Changing the rules: Ohio's Mahoning case shows rising dilemma
        of jobs vs. pollution." Wall Street Journal (July 27): 1, 22.

        Eight steel mills along the Mahoning River in Ohio were
        granted exemptions from EPA water pollution standards based on
        analysis suggesting the plants would close if forced to comply,
        taking about 20,000 jobs with them. While other steel com-
        panies have been able to comply, the Mahoning plants are out-
        dated, are energy- and labor-inefficient, and have rather small
        capacity. Since the plants are already marginal, the addi-
        tional large expenditures for pollution control would make them
        unviable. Local unions have been pressing the steel companies
        for promises that they will remain in operation once exempted
        from regulations but to date they have been unable to obtain
        even verbal commitments from the companies.

094
Kennedy, Edward M.
 1979 "Energy and jobs." Public Power 37 (November/December):
       34-35.

       Kennedy argues that past beliefs about the direct connec-
       tion between energy use and rate of growth in the economy are
       outmoded. Increased energy use is now associated with a weak,
       inflated economy and increased levels of unemployment. The
       author cites a study by the Public Resources Center of
       Washington, D.C., which indicates that implementation of a
       broad range of conservation and renewable energy systems would
       save $120 billion in energy costs and result in a net increase
       of 1.8 million jobs by the year 1990.

095
King, Jill A.
 1975 The Impact of Energy Price Increases on Low-Income Families.
       U.S. Federal Energy Administration, Office of Energy Informa-
       tion and Analysis. Available from National Technical Informa-
       tion Service as PB-262 098/7SL.

       Econometric analysis is presented on the effect of energy
       price increases on low-income families, using an energy data
       file for a national representative sample of 50,000 households
       in the continental U.S. Energy expenditures for each of six
       energy types were imputed for each household from this data
       file. Energy expenditures in 1974 were estimated using figures
       from a microsimulation which related energy consumption and
       disposable income for 1973 data. A substantial rise in house-
       hold energy expenditures occurred as a result of 1973-74 energy
       price increases. Households in New England and the Middle
       Atlantic regions were hardest hit. Although low-income house-
       holds spent less on energy and experienced smaller absolute
       increases in expenditures, these expenditures and increases
       represented a much larger proportion of disposable income than
       for high-income households, differing by a factor of 10.
       Single-family homes, larger families, and rural location were
       associated with larger impact because families with these
       characteristics use more energy. Impact of higher energy
       prices for home fuel expenditures did not vary by household
       characteristics, e.g., race, age, or occupation of head of
       household. Gasoline expenditures, and their impact, did not
       exhibit as wide a regional variation.

096
Krieger, Martin H.
 1970 "Six propositions on the poor and pollution." Policy Sciences
       1: 311-324.

       Krieger reports data indicating that public environmental
       problems will tend to be regressive in their impacts. The
       author indicated that the benefits of 23 existing federal envi-
       ronmental programs in the late 1960s tended to provide dispro-
       portionate benefits to affluent income classes. However, the
       data presented also suggest that a system of progressive
       taxation places disproportionate costs of these programs on the
       affluent as well.

1

097
Kruvant, William J.
   1975   "People, energy, and pollution."   In The American Energy
Consumer: A Report to the Energy Policy Project of the Ford
Foundation, ed.   Dorothy K. Newman and Dawn Day, pp. 125-167.
Cambridge, Mass.: Ballinger.

   Air pollution estimates for five metropolitan areas were
examined. A detailed study was then done of the relationship
between air pollution and socioeconomic characteristics of the
Washington, D.C., metropolitan area.   These studies yielded a
profile of the most likely victims of pollution. The Washington data show that socioeconomic characteristics associated
with disadvantage--poverty, occupations below management and
professional levels, low rent, and high concentrations of black
residents--go hand in hand with poor air quality. It is noted
that these groups produce little of the air pollution that
affects them.   The findings show that antipollution policies
have already helped disadvantaged groups, proving that well-
enforced policies can be effective.

098
Kurchak, John
   1978   "Solar energy: One union's viewpoint."   Alternatives   7
(Winter): 14-16.

   Kurchak, business manager of Local 285 of the Sheet Metal
Workers in Toronto, argues that solar energy represents the
possibility of increased employment for sheet metal workers.
President Carlough of the Sheet Metal Workers has testified
before the Canadian government that a modest commitment to
solar energy would employ all presently unemployed sheet metal
workers.   Kurchak believes that Canada should have a policy
promoting solar energy for employment reasons. A feasibility
study is cited which supports the practicality of this solar
policy. The author remarks on the failure of the Ontario and
federal governments to initiate solar policies.

099
Kutscher, Ronald E.
   1979   "The influence of energy on industry output and employment."
Monthly Labor Review 102 (December): 3-16.

   Several studies have attempted to predict the employment
and output impacts of increased energy prices.   This study
examines U.S. industry over the period 1958-1977 with the goal
of analyzing changes in employment and output occurring in the
1973-77 period when energy prices were rapidly escalating.
   The author compares output and employment in the most vs.
the least energy-, labor-, and capital-intensive industries.
He finds that, in general: (1) The output growth rate for the
least energy-intensive industries was greater than for the
general economy and much greater than for the most energy-
intensive industries, a reversal of long-term trends. (2) The
employment growth rates of the least and most energy-intensive
industries roughly correspond to the rates of growth in output

in those industries. (3) The most capital-intensive industries show, over all time periods, a considerably higher output growth rate than for the general economy. Since 1973, this difference has narrowed. The growth in employment in the most capital-intensive industries has always been below national average. Since 1973, employment in these industries has declined slightly. (4) The least capital-intensive industries have a lower output growth rate than the national average; absolute declines occurred in 1973-77. Employment growth in these industries has been lower than average, with declines occurring in 1973-77.

The author offers several caveats in interpreting the results of this study, primarily that the time period studied is relatively short. Given sufficient time to adjust, producers in energy-intensive industries may be able to change their capital stock to reflect high energy prices and scarcity. With this caveat in mind, it would appear that the U.S. economy is able to shift production away from energy-intensive production to production using relatively more labor.

100
Landsberg, Hans H., and Joseph M. Dukert
   1981 High Energy Costs--Uneven, Unfair, Unavoidable?   Baltimore: Johns Hopkins University Press.

The authors empirically confirm the notion that the poor have been hurt the most by energy price increases during the 1970s, although public assistance programs have softened the impact. The authors feel, however, that increased energy costs are necessary and unavoidable and that if the poor suffer disproportionately from increased energy prices, this problem is one of poverty and inequality and not of energy policy.

101
Lerner, L. J., and F. H. Posey
   1979 Comparative Effects of Energy Technologies on Employment. Sacramento, California: California Energy Commission.

This study analyzes the contribution to state and national employment of nuclear, coal, and oil power generation, the addition of insulation to existing buildings, and solar water heating. Direct, indirect, and induced employment effects are compared in terms of equivalent costs and with regard to energy produced or conserved. Every technology examined produces less employment per dollar spent compared to the average of national expenditures in general commerce. However, large variations among the technologies are evident. Solar heating and insulation provide more employment per dollar spent than does conventional energy generation. Power plant employment per kilowatt hour is greater than that of retrofit insulation, but solar energy water heating provides at least three times the employment per kilowatt hour than does any other plant type. This study confirms that investment in the least-cost alternative would help counter inflation and reduce unemployment.

102
Leung, K. C., and J. A. Klein
    1975   The Environmental Control Industry: An Analysis of Conditions
           and Prospects for the Pollution Control Equipment Industry.
           Report prepared for the U.S. Council on Environmental Quality,
           Washington, D.C., December.   Available from National Technical
           Information Service as PB-248 474/9SL.

           This detailed and comprehensive report focuses mainly on
           the sales and profit aspects of the pollution control equipment
           industry.   However, some more general impacts of environmental
           policy are offered, including impacts on employment.   Using
           mainly Bureau of Labor Statistics estimates of the employment
           generated per billion dollars of expenditures in pollution
           abatement sectors of the economy, the study estimates total
           employment impacts from total federal, private, state, and
           industry maintenance expenses of $15.7 billion in 1975.   This
           yields a figure of 1,000,000 direct and indirect jobs related
           to pollution control activities.

103
Levinson, Charles
    1975   "The malevolent workplace."   Ambio 4 (1):   24-9.

           This article provides a general discussion of toxic chemi-
           cals in the workplace, with emphasis on vinyl chloride.
           Levinson gives the history of evidence that vinyl chloride is a
           carcinogen; he documents the resistance of industry to health
           and safety standards, and the activities of unions and the
           International Labor Organization (ILO) in pressuring industry
           to adopt standards for vinyl chloride.
           Industry resistance is attributed to concern over profits,
           rather than to technical problems of monitoring and control as
           stated by the industry.   Levinson discusses the national and
           international role in promoting occupational health and
           safety.   He notes that countries with labor or socialist gov-
           ernments have experienced relative ease in passage of OSHA-type
           laws.   Levinson argues for (1) international labor organiza-
           tions such as ILO to coordinate the transfer of information on
           occupational hazards, and (2) industrial democracy as a means
           of promoting worker well-being in general and OSHA-type regula-
           tions in particular.   The article also contains a brief
           description of ILO programs for control of occupational cancer.

104
Levy, Girard, and Jennifer Field
    1980   Solar Energy Employment and Requirements, 1978-1985.   Prepared
           for U.S. Department of Energy, Office of Education, Business
           and Labor Affairs, and Office of Solar Applications.   Summary
           and Highlights available from National Technical Information
           Service as DOE/TIC-11154.

           This report describes in detail a survey conducted of over
           2,800 establishments engaged in all aspects of solar energy,
           from R & D to installation.   All types of active solar
           technologies were surveyed, and 1978 employment in solar

energy-related industry was determined, both in total and
disaggregated by type of application and by occupational
group. The authors then develop a methodology for projecting
future employment requirements based on employers' projections,
federal R & D projections, and solar energy market penetration
models.

Some of the major findings are as follows: (1) The turnover rate of solar firms is large. It is estimated that of the
2,800-plus establishments surveyed, only 2,000 were still in
existence in 1978. (2) Total employment requirements are projected to double by 1981 and to nearly triple by 1983. (3)
Over 40 percent of solar establishments have ten or fewer
employees. However, the 20 percent of establishments employing
over 400 persons accounted for over 80 percent of the solar
workforce. (4) It is estimated that 22,500 employees were
engaged in the solar industry in 1978. (5) Employees tend to
have moderate to high skill levels and training, although new
occupational specialties are rarely reported. A major short-
coming of this report is that it deals exclusively with active
solar applications, thus ignoring employment impacts of passive
systems.

105
Little, Arthur D., Inc.
  1972 The Economic Impact of the Pollution Abatement Equipment Indus-
        try. Report prepared for the U.S. Environmental Protection
        Agency, Washington, D.C., December.

        The authors make alternative assumptions about the level
of compliance with federal air and water pollution regulations
and translate these estimates into labor requirements in the
pollution abatement equipment industry. Assuming no federal
policy was in effect, the labor requirements would have been
21,000 jobs in 1972, 24,000 in 1975, and 31,000 in 1976.
Assuming a reasonable rate of compliance with the regulations,
employment in the pollution abatement industry was projected
to be 35,000, 49,000, and 75,000, respectively, for the same
years.

106
Little, Ronald L., and Stephen B. Lovejoy
  1979 "Energy development and local employment." Social Science
        Journal 16 (April): 27-49.

        Survey evidence from the Four Corners area of Utah near
the site of ongoing coal development is presented, showing that
residents of towns near additional development believe economic
benefits and improved job opportunities will result from the
proposed projects. However, evidence suggests that skills
necessary to participate in the development are largely absent
among local residents. Furthermore, survey evidence suggests
that residents would be unwilling to participate in training to
acquire the requisite skills. Projections based on past performance of locals in seeking and securing jobs in similar
energy projects suggest that of approximately 4,000 new jobs

available from the primary industry, only between 30 and 50 would be filled by local residents.

107
Logan, Rebecca, and Dorothy Nelkin
  1980 "Labor and nuclear power." Environment 22 (March): 6-13, 34.

The trade union movement in the U.S. is usually identified with a pronuclear position. However, significant antinuclear actions by unions have occurred. This article examines the economic and ideological roots of this difference in stance regarding nuclear power. The issue of safety vs. jobs is considered, as well as the implication of social vs. business unionism for labor's nuclear position. Opportunities for coalitions between certain sectors of labor and the antinuclear movements are assessed. The authors conclude that, ultimately, economic factors will mediate future union attitudes towards problems raised by nuclear power. However, a tradition of social unionism is not yet dead, so wider issues can still be expected to enter the debate from union quarters.

108
Lovejoy, Stephen B.
  1980 "Energy development and employment benefits: Who gets the jobs?" Ph.D. dissertation, Utah State University, Logan, Utah.

This thesis is an extended version of the paper by Ronald L. Little and Lovejoy.

109
Mason, Bert, and Keith Armington
  1978 Direct Labor Requirements for Select Solar Energy Technologies: A Review and Synthesis. Solar Energy Research Institute working paper. Available from National Technical Information Service as SERI/RR-53-045.

The object of this study is to synthesize several previous studies in order to estimate the gross labor requirements associated with probable levels of domestic solar energy implementation. Note that this labor requirement does not include losses from other energy sectors as solar use increases, nor does it include indirect employment effects resulting from dollar savings or losses associated with solar use.
The approach used is to synthesize three studies on the labor requirements of active solar systems to arrive at a best estimate of the labor input of a typical installation. Results of three studies of probable market penetration are synthesized and, when combined with labor requirements per installation, provide an estimate of probable total labor requirements.
The authors caution that the solar industry is immature, with little market experience, and that the studies are necessarily based in part on conjecture. They also alert the reader to methodological problems in several of the studies, including omission of several solar technologies from some studies and failure to project probable technological improvements.

110
McCallion, T.
  1976 "Solar energy: Potential powerhouse for jobs."   Worklife 1
       (August): 3-6.

       The employment implications of a fully developed solar
       heating and cooling industry are discussed briefly.

111
McCaull, Julian
  1974 "Energy and the workers."   Environment 16 (July/August):
       35-39.

       McCaull provides a summary of a meeting, sponsored by the
       Scientists Institute for Public Policy Information (SIPI),
       which brought together scientists, environmentalists, and union
       members (predominantly from a local of the Oil, Chemical, and
       Atomic Workers) to discuss issues related to energy and labor.
       McCaull argues that energy use has grown primarily because of
       increased industrial, rather than residential, use (with the
       exception of residential heating). Increased industrial energy
       use has been a result of general economic expansion and of
       individual firms' desires to increase their profit rates.
       Profit-seeking decisions by corporations have two major effects
       on the worker's environment: decreased maintenance of plant
       and equipment, in order to minimize production costs, which
       leads to increased worker danger; and increased automation,
       which results in a decrease in employment.
       McCaull argues that the emergence of the multinational
       corporation tends to make unions relatively powerless, since
       corporations can move to another country to avoid pressures
       brought by unions. International unions are needed to counter
       the power of multinational corporations. McCaull notes that
       the Oil, Chemical, and Atomic Workers are engaging in interna-
       tional union action. He stresses that prevailing economic
       forces will insure that present conditions continue unless
       unions and environmentalists bring pressure to bear on energy.

112
Middlebrooks, Joe
  1975 "Manpower needs of manufacturing industries."   Journal of the
       Water Pollution Control Federation 47 (December): 2850-2866.

       Middlebrooks uses an industry survey to estimate that
       575,000 work-years are used in operating the pollution control
       equipment in place as of 1975. This estimate appears to be
       about twice as large as other estimates. This discrepancy is
       perhaps partly explained by a tendency on the part of industry
       to overestimate the burden placed on them by pollution control
       regulations.

113
Miller, Alan S.
  1980 "Towards an environmental/labor coalition."   Environment 22
       (June): 3-39.

Miller summarizes the issues that traditionally have created antagonism between environmentalists and labor: a reduction in energy use, clean environment, and a workplace free of toxins are often seen as threatening jobs and as being highly inflationary. Miller cites several studies supportive of the notion that environmentally sound policies have not significantly affected employment or inflation. While the threat of job losses is blamed on environmentalists, Miller believes the real culprit is corporate failure to maintain efficient production technologies, coupled with a long-standing failure to protect the health and safety of workers and the public.

While the interests of environmentalists and labor are compatible, short-range contradictions in the aims of the two groups do exist. To overcome these obstacles, Miller suggests specific tasks aimed at building a coalition between the two groups. This coalition can then address the basic question of who should control America's wealth—corporate powers or the public?

114
Mitchell, Robert
   1979   "Silent spring/Solid majorities."   Public Opinion 2 (August-September):   16-20, 55.

Mitchell presents data from a national public opinion survey indicating that public support for environmental protection has not declined during the 1970s, as has often been asserted. Finding solid majorities of support for environmental control, even with respect to tough "trade-off" questions (e.g., in which respondents are asked whether they would support environmental protection if they would have to make economic sacrifices), Mitchell finds that public support for environmentalism is strong among virtually all major socio economic categories of the public. In particular, there is no evidence that blue-collar workers are significantly less supportive of environmental protection than white-collar workers.

115
Monroe, James, John Hill, James Jackson, and Walter Noiseaux
   1977   Jobs and Energy in New York State.   Report prepared for Legislative Commission on Energy Systems of the New York State Legislature, February 26.

This report provides comparisons of direct employment impacts of nuclear, coal, wind, and wood energy and three energy conservation approaches within New York State. Employment opportunities outside the state are excluded, and differences between conventional and alternative energy sources are exaggerated since conventional energy employment is largely outside the state. The following are summary data for a 1,000-megawatt plant or its equivalent:

Technology	Cost: $/kw	Jobs: 30-year cumulative totals
Nuclear	1000	12-14,000
Coal	700	12-13,000
Wind, without storage	600	18-19,000
Wind, with storage	1500	22-26,000
Wood, existing forests	900	65-80,000
Wood, plantations	900	60,000
Conservation	300	12,000
Conservation	600	24,000
Conservation	1000	41,000

116
Morrison, Denton E.
   1978 "Equity impacts of some major energy alternatives." In Energy
         Policy in the United States: Social and Behavioral
         Dimensions, ed. Seymour Warkov, pp. 164-193. New York:
         Praeger.

         Morrison assesses the probable distributional impacts of
   increases in the relative real price of energy (both direct,
   i.e., coal, oil, gasoline, electricity, and natural gas, and
   indirect, as a factor in the provision of other goods and
   services). Data are arrayed for 10 income classes by flows
   (monetary and energy) in 26 consumption categories. These data
   were recalculated from an input-output analysis of the 1960-61
   U.S. Bureau of Labor Statistics "Survey of Consumer Expenditures."
         The input-output analysis is unique in showing the energy
   impact of consumer expenditures, thus opening the way for a
   determination of the relative distribution of energy by class.
   The analysis is brought to bear on social equity considerations
   related to price increases, energy conservation, coal development, and nuclear development on persons, and to a lesser
   extent firms and communities, but especially on the poor.
         The analysis suggests that higher energy prices are
   regressive, particularly because the poor derive a larger
   proportion of their energy from direct forms. Energy conservation would entail transfer payments if high first-cost
   conserving technologies were to be made available to the poor.
   Moreover, the affluent would have to reduce their consumption
   disproportionately, especially in connection with indirect and
   nonbasic energy inputs. Nuclear and coal development could
   carry the seeds of inequitable risks and benefits. Inequity
   claims are briefly sketched in terms of quality of life. It is
   shown that the impact of energy on quality of life is more
   direct for the poor than for the affluent. A section is
   appended treating the equity impacts of six specific conservation strategies.

117
Morrison, Denton E., and Riley E. Dunlap
    1980 "Elitism, equity, and the environment." Paper presented at the
          annual meeting of the American Sociological Association, New
          York, August.

    Morrison and Dunlap argue that the issue of elitism in the
environmental   movement   must   be   disaggregated   into   three
distinct forms. Compositional elitism involves the accusation
that environmentalists are a privileged socioeconomic stratum.
Ideological elitism refers to the accusation that environmental
proposals are overtly intended to distribute benefits to environmentalists and/or costs to nonenvironmentalists, especially
the poor. Impact elitism involves the accusation that, inten-
tionally or not, environmental reforms have factually and
empirically distributed benefits to environmentalists and/or
costs to nonenvironmentalists, again especially to the less
privileged.
    The authors examine evidence relating to the three aspects
of elitism and suggest that the data tend to be ambiguous with
respect to each accusation.   While environmental movement
members are educationally and economically privileged, this is
no more the case than for other social movements (including
movements directly oriented toward enhancing social justice).
Moreover, among the general public, neither occupation nor
income tends to have a major correlation with proenvironmental
attitudes.
    With regard to ideological elitism, Morrison and Dunlap
see some elitist tendencies among prominent environmentalists
but demonstrate that environmental movement members and proenvironmental persons among the general public tend to be
politically liberal and supportive of the interests of the
poor. The impact elitism question is seen to be complex and
not well researched, although the authors lean toward an interpretation that the overall impacts of environmental reform have
a slight tendency to be regressive. However, Morrison and
Dunlap emphasize that the nature of these impacts varies
greatly from program to program and that definitive judgments
about environmental policy in general are not warranted.

118
Müller, Frank G.
    1980 "The employment effects of environmental policy: An inter-
          national comparison: Germany and USA."   Journal of Environ-
          mental Systems 10: 119-37.

    This paper reviews several studies done in the U.S. and
Germany on the employment impacts of environmental policy to
control air, noise, and water pollution. While the author
concludes that reliable estimates of employment impacts are
difficult to obtain, some tentative conclusions can be drawn.
    With current levels of expenditure for environmental pro-
tection, impact on employment is moderately positive. However,
expectations of significant improvements in unemployment levels
through environmental policy are unrealistic at the current
level of expenditure of about one percent of GNP in the U.S.

and Germany. Furthermore, in the U.S., these positive effects may erode when profit—oriented investment demands are offset by the loss in productivity and slower real growth associated with rising prices due to costs of pollution control. The author believes that current pollution control will not succeed in improving either long—run environmental quality or employment opportunities, since it is aimed at symptoms and not causes of environmental degradation.   Only when the economy is trans— formed to one based on renewable resources and strict conserva— tion of nonrenewable resources will society be equipped to adequately protect the environment and cope with unemployment and inflation.

119
National Petroleum Council
   1972 U.S. Energy Outlook.   Washington, D.C.: National Petroleum Council.

       This study was conducted to ascertain realistic approaches to the effective balancing of energy supply and demand beyond 1975.   The major options considered were: (1) increasing emphasis on development of domestic energy supplies, (2) much greater reliance on imports, and (3) restrictions on growth of energy demand.
       The study concludes that increasing the availability of domestic supplies is the best available option.   Accelerated development of domestic supplies would benefit all segments of society.   Unemployment would fall, individual income would rise, profit opportunities would improve, national security would be enhanced, and government revenues would increase. Greater reliance on imports involves uncertainties of supply, dependability, and price.   National security would be compromised and balance of trade problems would increase.
       The study argues that restrictions on energy use would prove costly, would alter lifestyles, would decrease consumer choice, and would cause unemployment.   These effects of decreased energy use remain assumptions throughout the study, with no supporting evidence offered.   The bulk of the study is an analysis of feasible alternatives of meeting an assumed 4.2 percent annual growth in energy use.

120
National Technical Information Service
   1980 Labor and Energy Impacts of Energy—Conservation Measures. DOE/CS/20258—2.   Springfield, Va.:   National Technical Information Service.

       Three papers are presented discussing labor and energy impacts of three energy conservation measures.   Using input—output analysis, Robert Herendeen examines the labor and energy implications of improving the thermal integrity of homes.   His analysis shows that this program could save 0.5 quads per year in 1990 and generate 30,000 additional jobs. Salary distribution effects are also examined.   Carole Green develops and presents industry—occupation wage matrices necessary to analyze employment effects.   John Nangle presents another analysis of the labor and energy effects of producing new homes with better thermal integrity. His results are similar to Herendeen's.

121
Neuhaus, Richard
    1971 In Defense of People: Ecology and the Seduction of Radical-
          ism. New York: Macmillan.

Neuhaus's book is commonly regarded as the first major critique of the environmental movement from the vantage point of social justice and remains one of the most vitriolic indictments of the tendency of environmentalists to focus on superficial problems and to ignore the concerns of those people—the poor—who suffer from both economic inequality and environmental problems. The basic argument of the book is that ecology, while not inherently in contradiction with the interests of the poor, has represented a diversion of attention from the immediate survival interests of the poor and thus is, as indicated in the book's subtitle, a "seduction of radicalism."

Moreover, Neuhaus feels that many environmentalists have tended to assume an antihuman posture, seeing, for example, that "too many people" are destroying the planet without considering how political and economic inequality has led to population problems. Neuhaus is especially concerned that the growth of environmental sentiment will be accompanied by coercive government policies at home and abroad which will have the effect of locking the underprivileged in their subordinate positions in order to protect ecosystems.

122
New York Times
    1976 "Workers outshout demonstrators at nuclear plant." New York
          Times, March 1: 38.

This article describes a confrontation between union members involved in construction of the Indian Point nuclear reactor in New York State and antinuclear demonstrators. Then-prospective U.S. Senate member Bella Abzug spoke at the rally, calling for investigation of nuclear safety, for income protections for workers that might be displaced by nuclear shutdowns, and for a coalition between workers and environmentalists. Heckling from union members largely drowned out antinuclear speakers.

123
New York Times
    1979 "Ohio air pollution rules are relaxed to save jobs." New York
          Times, June 7: A18.

EPA regulations specify different levels of pollutants that can be emitted from coal-burning utility plants, depending on whether the plant is in a rural or urban location. Cleveland plants are in violation of urban standards as a result of burning dirty Ohio coal in the absence of pollution control equipment to clean up the emissions. To comply with regulations, the plant would have to either install scrubbers or burn clean western coal. The first option can be forced under EPA law to protect local jobs but would result in unpopular rate increases. The second option would result in an increase in coal mining jobs in western states but a loss of about 5,375 mining and related jobs in Ohio. To avoid conflicts associated with either option, the Cleveland plants were declared rural since they are located on a lake shore. This action allows higher levels of emissions. Environmentalists oppose this action on legal grounds. They claim political concerns were the sole motivation for the change in status and that the move cannot be justified on legal grounds.

124
Newman, Dorothy K., and Dawn Day
    1975 "The energy gap: Poor to well off." In The American Energy Consumer: A Report to the Energy Policy Project of the Ford Foundation, ed. Dorothy K. Newman and Dawn Day, pp. 87-124. Cambridge, Mass.: Ballinger.

    A description is presented of how poor, middle-income, and well-off families use energy, based on data from the Washington Center for Metropolitan Studies' Lifestyles and Energy Surveys, conducted May-June 1973 (household interviews) and June-September 1973 (acquisition of billing data from utilities) on a nationwide multistage area probability sample (N=1,455) of heads of households. It was found that the poor use less energy, pay relatively more for the energy they must have, and, more than any other American group, suffer from exposure to the residuals of energy production and consumption. The energy gaps were found to be greatest in gasoline use.

125
Not Man Apart
    1977 "Jack Mundey speaks: 12 Australian success stories." Not Man Apart 7 (March): 1-3.

    Jack Mundey, past president of the Australian Builders-Laborers Federation, details some of ABLF's past involvement in social issues. In the early 1970s, ABLF became involved in environmental action when it noticed that there was a high rate of highrise construction despite shortages of housing, hospitals, and schools. The union is involved in so-called "green bans"--strikes against projects felt to be environmentally inappropriate. Green bans have been imposed on 42 projects over four years. Bans are imposed only with strong community support, which has often been requested by community groups. The aim of green bans is usually to give the community a voice in the development of projects, rather than merely shutting down projects. To this end, alternatives to the original plan are often developed--e.g., working-class flats instead of capital-intensive, labor-extensive highrises.

126
Olsen, Marvin E.
    1978 "Public acceptance of energy conservation." In Energy Policy in the United States, ed. S. Warkov, pp. 91-109. New York: Praeger.

Olsen reviews the previous literature on factors that lead to energy conservation among the American public. He points out that energy price increases tend to be relatively ineffective and inegalitarian as a rationing mechanism. Price increases are ineffective because the poor have little room to conserve from their already low levels of energy consumption, while the affluent are relatively impervious to price increases. The bulk of conservation has been accounted for by middle-income families, while at the same time exacting the greatest hardships on the poor. The author argues that effective, egalitarian energy conservation strategies will necessarily involve public intervention in the energy marketplace, which will undoubtedly be resisted on political and economic grounds by the affluent.

127
Paehlke, Robert
    1979 "Occupational health policy in Canada." In Ecology versus
          Politics in Canada, ed. W. Leiss, pp. 97-129. Toronto:
          University of Toronto Press.

Beginning with a case study of the health effects of asbestos, Paehlke discusses the nature of workplace health hazards resulting from toxic chemicals and substances. He argues that the most dangerous forms of bodily response to hazardous substances tend to have delayed effects, making it possible for industries to cause serious health problems among their workers without any possibility of immediate detection. Paehlke argues that industry's scientific research in exploring the potential health hazards is extremely suspect, given the experience with asbestos and other substances; for example, industry will usually report data from epidemiological tests of whole industries or occupations, thereby concealing the fact that a few workers perform particularly hazardous tasks. The author argues that the power of industry and inattention of the labor movement to occupational health are largely responsible for these problems.
    Paehlke provides a set of policy recommendations for unions to deal with occupational health, including revisions of national medical recordkeeping techniques, applying political pressure for the passage of comprehensive "right-to-know" legislation about workplace hazards for individual workers, union scrutiny of and independent research on workplace epidemiology, union demands for union-hired, employer-paid occupational health technicians to do ongoing health tests among all workers, insistence that all new substances should be thoroughly tested prior to introduction into nonlaboratory workplaces, and several others.

128
Perlman, Robert, and Roland Warren
    1977 Families in the Energy Crisis: Impacts and Implications for
          Theory and Policy. Cambridge, Mass.: Ballinger.

A multistage area probability sample (N=1,440) was conducted in November 1974 in Hartford, Connecticut (N=658),

Mobile, Alabama (N=483), and Salem, Oregon (N=243), to determine the impact of energy problems on households which differ
by income level, social characteristics, geographic region, and
access to energy supply. A follow-up questionnaire was mailed
in November 1975 to ascertain changes in both behavior and
attitudes since November 1974.

The study indicates that self-serving economic and political motives were blamed by a majority of respondents for artificially generating the crisis.   Some perceived it as a
noncrisis, others saw it as potentially threatening.   Adjustments were made by very sizable numbers of families in their
use of the car.   Considerable proportions of households cut
back on the conveniences of home heating, air conditioning, and
the use of electricity.   Vital areas like health, well-being,
and the basic financial condition of families, taken as a
whole, did not seem affected.   Most adversely impacted by the
crisis were families whose wage-earners were forced to leave
their jobs.   The viability and life-styles of families other
than these did not appear to be seriously affected.

Aggregate energy savings were found to be 12 percent
before attrition set in.   With regard to equity impacts,
"qualitatively and quantitatively different resources had the
effect of setting differential constraints and opportunities
within which families had to adjust to the energy crisis and
had to absorb its costs and burdens."   Higher-income families
demonstrated greater flexibility in absorbing cost increases.
Poor families cut back more as a percentage of their precrisis
consumption and endured a heavier burden of unemployment. Also
discussed are the 1977 energy crisis and public policy on
energy conservation.

129
Peskin, Henry M.
   1978   "Environmental policy and the distribution of benefits and
           costs."   In Current Issues in U.S. Environmental Policy, ed.
           P. R. Portney, et al., pp. 144–163.   Baltimore: Johns Hopkins
           University Press.

           Peskin notes that the predominant distributional conse-
           quence of environmental policy is interregional rather than
           according to economic class. He notes, for example, that the
           major beneficiaries of air pollution control policies are the
           five "dirtiest" (all Northeastern) Standard Metropolitan
           Statistical Areas, while greatest costs are borne by residents
           of areas with few low-income families and high per family auto-
           mobile ownership.   Net benefits are distributed in a manner
           similar to gross benefits, with residents of the Jersey City,
           New York, Erie, Newark, Detroit, Paterson, and Chicago SMSAs
           showing net benefits, in dollars per family, of $300 or more.
           With regard to distributional benefits and costs by income
           and racial groups, Peskin notes that, overall, air quality
           improvement is not strongly progressive or regressive.   How-
           ever, the effects of air quality improvement exhibit dramatic
           distributional variations when disaggregated into industrial
           pollution control and regulation of automobile emissions.
           Industrial pollution control tends to have a progressive dis-

tribution of net benefits, while regulation of automobile
emissions has regressive distributional impacts.

130
Pledger, Normal
    1978 "Energy development, jobs and the environment."   Paper pre-
          sented at American Nuclear Society's Second Conference on
          Environmental Aspects of Non-Conventional Energy Resources,
          Denver, Colorado, September 26-29.

          This paper urges development of U.S. energy policies that
          (1) emphasize conservation and alternative energy sources as
          well as increased production of fossil fuels; (2) recognize the
          potential for energy development to ease unemployment rates;
          and (3) recognize the need to protect the environment as energy
          resource development proceeds.

131
Portney, Paul R.
    1976 The Distribution of Pollution Control Costs: A Literature
          Review and Research Agenda. Report prepared for the Panel on
          the Sources and Control Techniques of the Environmental Re-
          search and Assessment Committee, National Academy of Sciences,
          Washington, D.C., February.

          Portney provides a comprehensive assessment of the early
          1970s literature on the distributional effects of environmental
          programs. The results generally suggest that the poor bear
          proportionately larger burdens of environmental protection due
          to taxation and consumption effects.

132
Renshaw, Edward F.
    1981 "Energy efficiency and the slump in labor productivity in the
          U.S.A." Energy Economics 3 (January): 36-42.

          Energy consumption was included in an aggregate Cobb-
          Douglas production function for the U.S.A. in order to analyze
          the impact of recent changes in energy consumption on labor
          productivity and GNP. Since the 1973 oil embargo, labor pro-
          ductivity has declined dramatically. This slump has been
          attributed to higher energy prices having induced substitution
          of labor for energy. The analysis shows that reductions in
          energy use have not reduced real GNP by more than 0.3 percent
          and that only about 20 percent of the recent slump in labor
          productivity can be attributed to higher energy prices.
          However, the source of the remaining decline in productivity is
          unresolved. The author concludes that the slump should not be
          viewed with alarm, since it may aid goals of full employment
          without further increases in energy consumption.

133
Rodberg, Leonard S.
    1979 Employment Impact of the Solar Transition. Study prepared for
          the Subcommittee on Energy of the Joint Economic Committee.
          Washington, D.C.: U.S. Government Printing Office.

Two hypotheses are made in this report. First, it is hypothesized that it is possible to produce the same quantity of goods and services with less energy than is presently used and to achieve increased GNP through conservation and renewable energy resources. Second, it is hypothesized that these changes—conservation and the use of renewable energy resources—can lead to an increase in employment.

Rodberg notes that "business as usual" projections indicate continued difficulty in achieving full employment. Increased direct and indirect consumer spending on conservation and a switch to alternative energy sources are essential strategies for achieving full employment in the U.S. The report suggests that these changes could be implemented by spending $65.6 billion per year, which represents 13 percent of the Bureau of Labor Statistics' estimate of private spending projected through 1990. The author suggests that such a program would decrease energy consumption in the U.S. by 44.9 quads, compared to standard projections of energy consumption, leaving energy consumption essentially unchanged from 1977 levels. One-half of this reduction in energy consumption would derive from energy conservation and the other half from utilization of solar energy.

The suggested level of spending on conservation and solar would lead to the following job accounting:

2,170,000	direct jobs gained in conservation industry
1,137,000	direct jobs lost in nonrenewable energy industry
1,870,000	indirect jobs gained through increased spending power made available through conservation
2,903,000	net increase in jobs over the "business as usual" future

Rodberg recognizes that there are major barriers to the implementation of this strategy. Individuals who face high initial costs must install conservation and solar equipment. These individuals will face the high cost of borrowing so that an alternative financing plan sponsored by the federal government may be necessary.

The author estimates that solar energy could account for 10.2 quads, or 13 percent of 1990 energy consumption. Compared to BLS projections of energy expenditures in 1990, the use of solar energy would result in gross savings of $118.8 billion per year. After deducting the spending of $65.6 billion per year for conservation and solar equipment, the net savings from use of solar power would be $53.2 billion per year.

134
Rost, R. F.
   1978 "Pollution controls and labor productivity growth." Ph.D. dissertation, Northwestern University, Evanston, Illinois.

This study analyzes the slump in labor productivity between 1967 and 1975 and attempts to estimate the extent to which pollution control expenses incurred in this period contributed to the slowdown. Production functions are constructed for the 11 major domestic industries in the accounts of the

Bureau of Economic Analysis and for most two-digit manufacturing industries. For the four industries incurring heaviest
pollution control expenses (mining, manufacturing, utilities,
and transportation), detailed series that included both pollution control and other capital requirements among the independent variables were prepared. For other industries, capital
inputs were ignored because this information is generally
unavailable. The analysis reveals that a break in the trend of
increased productivity growth occurs in the 1967-1975 period.
The cyclically corrected slowdown was most severe in construction, with a drop of 7.5 percent, followed by mining with a 5.3
percent reduction in the rate of growth of productivity.

135
Rostow, W. W.
    1978 "Energy, full employment, and regional development."   Paper
          presented at the American Association for the Advancement of
          Science, Washington, D.C., February 12-17.

              This paper discusses the ways national and regional econo-
          mies of the U.S. would be affected by a successful implementa-
          tion of the 1977 National Energy Plan. Historical evidence on
          the shifts in investment that would occur and the effects of
          these shifts are considered. The extent to which existing
          models that attempt to relate energy consumption and macroeco-
          nomic performance are relevant to the plan's implementation are
          discussed. Some possible regional applications of the analysis
          are offered. Results of the analysis are presented in detail.

136
Schacter, Meg
    1979 "Addressing employment effects of solar energy adoption."
          Paper presented at ISES-AS/ET AL Solar Action at the Local
          Level Conference, May 28-31.

              This paper details expected impacts of solar heating on
          employment. The author argues that solar systems can produce
          two to eight times more direct jobs than conventional power
          plants.

137
Schnaiberg, Allan
    1975 "Social syntheses of the societal-environmental dialectic: The
          role of distributional impacts." Social Science Quarterly 56
          (1): 5-20.

              This article summarizes existing empirical research on the
          distributional consequences of the energy crisis. The energy
          crisis was found to have had net regressive distributional
          impacts. For example, the poor suffered more than the well-
          to-do in terms of income loss, unemployment, and impacts on
          life-styles; small businesses were hurt more than large
          corporations; and the Nixon administration was able to use the
          energy crisis to justify curtailing "nonessential" federal
          governmental expenditures in such areas as health, education,
          and welfare.

138
Schnaiberg, Allan
    1980 "Saving the environment: From whom and for whom?"   Paper
          presented at the annual meeting of the American Sociological
          Association, New York, August.

          Recognizing that "appropriate" or "soft" energy technology
          has the potential for reducing demand for inanimate energy and
          for increasing employment levels, Schnaiberg is critical of
          many appropriate technology (A.T.) proponents for their anti-
          labor ideologies and their naive views of the American class
          structure. Schnaiberg sees the dominant ideology of the A.T.
          movement to be an essentially conservative, individualistic,
          private-property-oriented, decentralist "survivalism" which
          tends to be hostile toward those, such as the poor, who are
          dependent upon the state for their economic survival. Seeing
          this ideology in many respects indistinguishable from small-
          town, Midwestern Republicanism, the author suggests that this
          ideology is impractical in a complex political economy that is
          premised on massive state intervention and on control of
          investment decisions by a small group of propertied persons in
          large corporations and investment houses.
          Schnaiberg also argues that the A.T. movement's prescrip-
          tion for a soft-energy future leading to more employment is
          flawed by the fact that jobs created by deployment of appropri-
          ate technology will tend to be low-skill (and presumably
          low-wage) jobs. He suggests that proponents of appropriate
          technology tend to have a naive view of class structure and
          that welfare issues cannot be solved merely through the labor-
          intensity of employment in a soft-energy economy. Given the
          risks and uncertainties of a shift toward soft-energy technolo-
          gies, Schnaiberg is not sanguine about a substantial fraction
          of labor being willing to cast their lot with a scenario
          promising low-skill jobs and the dismantling of public services
          provided by a centralized state.

139
Schnaiberg, Allan
    1980 The Environment: From Surplus to Scarcity.   New York: Oxford
          University Press.

          Chapter V, "The Expansion of Production: Capital, Labor,
          and State Roles," discusses the role of organized and nonunion-
          ized labor in the pattern of post-World War II economic expan-
          sion and environmental destruction in the U.S.   Schnaiberg
          observes that U.S. unions tend to be more "economistic" in
          their strategies than Western European unions and that U.S.
          unions have primarily emphasized increased wages while Western
          European unions have taken a greater interest in national
          economic planning, codetermination of production decisions, and
          improvement of working conditions.
          Schnaiberg argues that major corporations and labor unions
          have entered into a de facto economic growth coalition: unions
          give up their ability to influence corporate technology and
          production strategies in exchange for high wages, while corpo-
          rations, given their rapid postwar expansion and growth,

tolerate high wage demands to preserve their prerogatives to automate the workplace and to subject workers to industrial hazards. Schnaiberg notes that this growth coalition may tend to dissolve if economic stagflation undermines the ability of state decision makers and corporate officials to generate sufficient economic growth to make the postwar bargain viable. He suggests, however, that the status of unorganized workers has tended to deteriorate during the 1970s for two major reasons. These workers do not have unions to represent their interests, and the forms of technology adopted by corporations tend to be capital-intensive and labor-extensive, hence undermining the low-skill jobs that unorganized workers usually occupy.

140
Schnaiberg, Allan
    1981 "Energy, equity and austerity: Some political impacts of energy rationing by price."   Paper presented at the annual meeting of the Society for the Study of Social Problems, Toronto, August.

        Schnaiberg notes that energy price increases have represented a de facto system of rationing which has led to the forced conservation of energy. This paper explores several political implications of energy rationing by price. The author argues that the impacts of energy price increases (e.g., truncated geographical mobility due to higher transportation expense and the general loss of discretionary income because of energy-induced inflation) have led to a "rightward" turn in economic outlook among consumers.
        Schnaiberg stresses that this rightward turn among government bureaucrats regarding energy-related economic problems helped to lead to a new rightist political coalition. This new rightist coalition combines the localistic Proposition 13ism of Howard Jarvis with the interests of industrial elites. Schnaiberg suggests that the new conservative coalition has been able to mobilize consumers pinched by rising energy prices and growing inflation to support reduction of the level of taxation. Reducing the level of taxation has become translated into dismantling post-New Deal transfer programs, which has, of course, dramatically affected the poor. "Class conflict" has thus been shifted from the more traditional form of conflict over the distribution of public funds to the poor to one of "consumers" versus "the government"--a pattern of conflict which the author feels will work against the interests of workers and the poor.

141
Schreker, Ted
    1975 "Labour and environment: Alternatives conference report." Alternatives 4 (Winter): 34-42.

        Schreker reports on a conference sponsored by Alternatives and attended by union members, union officials, and environmentalists from government, university, and citizen groups. Schreker observes that the "jobs vs. environment" issue is often an industry-sponsored "red herring." Unionists at the conference were concerned with the effects of pollution on society as a whole, not just workers as a group, although Schreker wonders if this attitude is shared by the union rank-and-file. Workplace contaminants are seen as an obvious area of joint action by unionists and environmentalists. Schreker notes the inadequacy of workplace safety laws and of the enforcement of existing laws.

Conference participants agreed that workers should not bear the burden of increased costs due to environmental protection and workplace protection (e.g., through the loss of jobs due to increased costs or through wage decreases to compensate for these increased costs). Free enterprise and the profit motive are seen as the driving forces behind environmental problems and workplace hazards.

142
Severo, Richard
   1976   "Environmentalists call jobs a false issue." New York Times, March 1: 1, 38.

Environmentalists are more vocal about their concerns than ever before. However, they also face greater opposition from government and industry as well as facing severe funding problems. While the bulk of this article discusses financial difficulties of environmental groups in and around New York City, there is some discussion of environmentalists being scapegoated for wider economic problems. Severo claims that, to the contrary, environmental controls create rather than destroy jobs.

143
Smeltzer, K. K.
   1980   Employment Impacts of Selected Solar and Conventional Energy Systems: A Framework for Comparisons and Preliminary Findings. ANL/EES-TM-116. Springfield, Virginia: National Technical Information Service, January.

Smeltzer presents a framework for estimating various effects of energy systems—direct, indirect induced, displacement, disposable income, and qualitative employment. The methodology is applied to selected solar energy and conventional energy systems. The analysis suggests that solar energy provides higher direct and indirect employment than does conventional energy. Induced employment effects caused by changes in disposable income of consumers are also shown to be highly significant. Smeltzer also argues that dispersed solar energy projects have a more beneficial impact on communities regardless of the employment impacts, largely because the smaller scale does not overly strain the assimilative capacities of small communities.

144
Smeltzer, K. K., and D. J. Santini
   1979   "Employment from solar energy: A bright but partly cloudy future." Paper presented at the First Annual Community Renewable Energy Systems Conference, Boulder, Colo., August 20.

This study synthesizes several previous studies in an attempt to estimate employment needs of solar vs. conventional energy systems. The authors caution that making comparisons between an industry with a well developed data base regarding employment (conventional systems) and one without such a reliable data base (solar) is hazardous; they advise caution in interpreting the results.

The comparisons are based on employee-years per BTU delivered per year. Depending on the type of technology considered, combined direct and indirect employment appears to be one to five times higher for solar than for conventional energy systems. Net employment impact—i.e., increases due to solar less decreases in conventional systems—appear positive, although the specific results are not reported. Also not reported, although discussed as important in the article, are the respending effects of dollar savings on employment opportunities if solar is less expensive than conventional systems.

145
Smith, James Noel, ed.
   1974 Environmental Quality and Social Justice in Urban America. Washington, D.C.: The Conservation Foundation.

This monograph represents the proceedings of a conference devoted to beginning a dialogue among conservation groups, organized labor, and other "social justice" groups. The context for this conference and monograph was the substantial degree of criticism of the alleged elitism of environmentalists put forth by persons such as the Rev. Richard Neuhaus, one of the conference participants and authors in this collection.

The contributors to this volume are generally sympathetic with the goal of uniting the movements for environmental quality and social justice, recognizing that environmental quality and social justice can be largely compatible and that environmentalists and labor have a common interest in preventing "environmental blackmail" by industry.

The representatives from labor, environmental groups, and academia all tended to agree that the environmental movement has given inadequate attention to problems of the inner-city environment and of inner-city residents and that environmentalists tend to retain an overly rural conception of environmental problems (e.g., focusing on preservation of open space and wilderness). The various contributors provide a number of useful suggestions for blending the interests of environmental quality and social justice, many of which are relevant to organized labor.

146
Solar Energy Research Institute
   1979 Organized Labor and Solar Energy. Available from National Technical Information Service as SERI/TR-62-148.

This SERI report is based on a workshop among leaders of 13 unions, along with DOE and SERI officials, regarding solar energy, which was followed by in-depth interviews with leaders

of 18 interested unions, by field visits to 22 selected building and trade councils, and by a survey sent to 100 selected
building and trade councils. The purposes of the project were
to develop channels of communication between organized labor
and DOE-SERI; to identify programs, policies, and structures of
which energy officials should be cognizant; and to determine
present union orientations toward solar energy and the degree
of future union involvement in solar energy.

The findings of the project include:

(1) Many unions are willing to participate in solar technology and have most of the requisite skills.

(2) The relatively simple technology of solar energy is
not reflected in the technical literature, and the perceived
complexity of solar energy scares off many potential customers.

(3) Few jurisdictional problems exist in the development
of solar energy, and those problems which do exist are
receiving attention now.

(4) Reliability has been seen as an obstacle to the
development of solar energy, but reliability problems can best
be overcome through installation of solar equipment by union
craftsmen.   "The best solar is union solar," in the words of
Edward Carlough, president of the Sheet Metal Workers.

(5) SERI should have a major role in development and dis-
semination of information about solar energy.

(6) Some unions are already involved in solar energy, but
in a minor way.

(7) Some unions feel that they should play a role in
influencing their members to become consumers of solar energy.

(8) Most, but not all, union representatives contacted
feel that solar energy has a major role to play in the U.S.
energy future and that their union should have a role in solar
energy development, either short- or long-term.

147
Solow, Robert
   1972 "The economics of pollution control." In Jobs and the Environ-
         ment: Three Papers. Berkeley: Institute of Industrial Rela-
         tions, University of California.

         Effective pollution control need not reduce employment,
   since 2.5 percent of GNP could be allocated to pollution control without major employment declines. Solow suggests either
   taxes or effluent charges as sources of finance for pollution
   abatement.

         Effluent charges are claimed to be as or more effective
   than subsidies or direct control because: (1) these charges
   would correct the flaw in pricing policy by providing incentives to internalize externalities of pollution, and (2) effluent charges generate revenue that can be used for any social
   policy, including amelioration of employment dislocations.
   Solow argues that environmental control is seen as costly
   because these costs are easily seen and measured while environmental benefits, although substantial, are not readily apparent
   or easily measured.

148
Stanfield, G. G.
  1978 Predicting Socioeconomic Impacts of Energy Development: Net
        Changes in Employment in Response to Increases in Basic Employ-
        ment. Working Paper No. 44. University Park, Pennsylvania:
        Pennsylvania State University, May.

          Expansion of industrial activities into rural areas has
        led to interest in developing methodologies for predicting the
        indirect and induced employment effects of these activities.
        Stanfield reviews existing research in this area which would be
        of use to individuals interested in making socioeconomic
        predictions as part of environmental impact statements. The
        work emphasizes the difference between structural ratios and
        multiplier coefficients. Factors included in the discussion
        include population size, length of study period, and population
        density. The specific conclusions of this study are reported
        in four guidelines for predicting employment multiplier
        effects.

149
Starr, Chauncey, and Stanford Field
  1979 "Economic growth, employment and energy." Energy Policy 7
        (March): 12-22.

          The authors examine past trends (late 1800s to the
        present) and conclude that increased labor productivity,
        requiring increased energy and capital inputs, has resulted in
        substantial social progress and an increase in the standard of
        living. If a continuation of this progress is desired, the
        authors state that a continuation of the trend of increasing
        energy intensity is necessary. Energy consumption for the year
        2000 is projected to be 150 quads per year if energy con-
        straints on improvement in social well-being are to be avoided.
        The authors suggest that this energy be derived primarily from
        coal and uranium, with gas and oil reserved for nonsubstitut-
        able uses.
          This study is of limited usefulness, since it is based
        entirely on past trends, does not incorporate known potential
        for substitutability between energy and labor, and assumes very
        modest conservation potential.

150
Stephen Soboyka and Co. and McKee-Berger-Mansuets Inc.
  1972 The Economic Impact on the Construction Industry of Additional
        Demands Caused by New Environmental Protection Standards.
        Report prepared for the U.S. Environmental Protection Agency,
        Washington, D.C., December.

          This study estimates the impact of pollution controls on
        the demand for labor. The methodology used was to compare the
        labor requirements for construction, first assuming no federal
        regulations and then assuming a reasonable rate of compliance.
        The study concludes that an estimated 35,000 additional
        construction jobs would be created in 1976 due to pollution
        control regulations.

151
Stretton, Hugh
    1976 Capitalism, Socialism, and the Environment.   Cambridge:
         Cambridge University Press.

         Stretton argues that advanced industrial societies can
    adapt to problems of scarcity and environmental destruction
    along three very different political paths or options: conser-
    vative, liberal, or socialist.   The conservative path would
    involve a coerced squeezing of the consumption of workers in
    order to maintain supplies of raw materials to industry, and
    would be sharply inegalitarian.   The liberal option ("business
    as usual") would involve a gradual increase in inequality as
    environmentally induced stagflation ate away at the incomes and
    employment security of workers.
         Stretton argues that a socialist form of adaptation to
    scarcity is in the best interest of workers, although his image
    of socialism lies more toward aggressive social democracy than
    toward state ownership of all productive and consumptive
    property.   This socialist option would primarily involve
    squeezing the energy- and materials-intensive consumption of
    the affluent and "institutionalizing" (i.e., nationalizing)
    industries only when there is a disjunction between private
    profit and the common good.   In particular, Stretton stresses
    the need to retain "domestic" property (i.e., homes, gardens,
    autos) in private ownership, since collective ownership of such
    property tends to stifle the productivity of the household
    economy.

152
Stutz, John, and Paul Christensen
    1979 "Jobs and energy: Current results and future directions."
         Paper presented at ISES-AS/ET AL Solar Action at the Local
         Level Conference, May 28-31.

         Results of employment forecasting models are presented
    which predict the employment impacts associated with solar
    scenarios for New York and Massachusetts.   The analysis pre-
    dicts a net gain in employment for New York and a smaller gain
    for Massachusetts.   The article urges further research on the
    application of such models to district heating.

153
Tinker, Jon
    1974 "Tin hatted conservationist."   New Scientist 64 (June 6):
         620-24.

         This article provides more information about Jack Mundey,
    elaborating upon his general political views (see Not Man
    Apart 7 (5): 1-3, 1977).   It is noted that Mundey is a member
    of the Australian Communist Party.

154
Turner, J. C.
    1978 "National energy policy and U.S. labor."   Paper presented at
         the National Energy Forum VI, Washington, D.C., May 17.

Turner suggests that organized labor would profit from the maximization of domestically produced conventional and nuclear energy. The U.S must, he feels, encourage the development of known domestic sources of energy that can be implemented with known technologies. Government regulations must be streamlined to remove impediments to this development; otherwise, energy shortages will result in unemployment as the economy contracts.

155
U.S. Congress, House Committee on Banking and Currency
    1974 Energy Security and the Domestic Economy: Impact on Prices, Employment and Consumption. Washington, D.C.: U.S. Government Printing Office.

The portions of this study relevant to the employment effects of the energy situation revolve around the "structure of employment" (i.e., the percent of total output accounted for by particular industries) and the "organization of employment" (the relative intensities of capital, labor, and energy in production). It is noted that the structure of employment will change as industry shifts toward more labor-intensive production. Permanent unemployment can occur if the skill levels of expanding and contracting labor markets differ, or if these shifts involve different geographic locations, or if they occur suddenly and without warning.
It is unclear if the organization of employment will shift as a result of changes in energy prices, since the evidence is contradictory on the question of whether capital and energy are substitutes or complements. If energy and capital are substitutes, increasing energy costs lead to declining energy use and to increasing capital intensity and associated declining labor intensity. If energy and capital are complements, increasing energy costs result in declining energy and capital intensity, with increasing labor intensity. The report cites one study supporting each position. Nevertheless, the report concludes that the effects of energy shortages on the structure and organization of employment will likely be quite small when averaged over the entire economy.

156
U.S. Congress, Joint Economic Committee
    1978 Creating Jobs Through Energy Policy. Hearings Before the Subcommittee on Energy. Ninety-Fifth Congress, Second Session, March 15-16, 1978. Washington, D.C.: U.S. Government Printing Office.

Witnesses representing federal, state, and local governments; labor; academia; and minority groups appeared before the committee. All urged coordination of energy and employment policy. Specific proposals on the nature of that coordination varied considerably. One fairly consistent theme was a call for "employment impact assessments" for proposed energy policies and federally supported energy projects. This idea sprang from the widely recognized fact that labor requirements of various energy alternatives are very different.

Other testimony noted that: (1) The current tax structure subsidizes capital, creating incentives for capital-intensive as opposed to labor-intensive development. (2) Since conservation and renewable energy sources are more labor-intensive than conventional solutions, these options should be promoted to decrease unemployment. (3) Current energy problems tend to disproportionately affect the income and employment situations of low income groups; consequently, energy policy should be designed to shield these groups. (4) Some witnesses held that despite conservation, increasing energy consumption is necessary for stable or increasing employment opportunities, while others felt that decreasing energy use is not inconsistent with employment goals. (5) Most witnesses felt that increasing GNP was necessary for maintenance of employment standards.

157
U.S. Department of Commerce, Bureau of the Census
    1980   Environmental Quality Control--Government Finances: Fiscal
           Year 1977-78.   State and Local Government Special Studies #97.
           Washington, D.C.: U.S. Government Printing Office.

           This series is published every year and reports the results of federal, state, and local surveys of environmental control programs.
           The 1977-78 programs accounted for $11.9 billion in public spending for environmental protection. Unfortunately, detailed employment results were discontinued after earlier editions. Employment figures are currently given for local governments only.

158
U.S. Department of Energy, Office of Policy and Evaluation
    1979   Creating Jobs Through Energy Policy: A Guide to Resources for
           Decision Makers.   Available on microfiche only from National
           Technical Information Service as DOE/PE-0013.

           This handbook presents basic economic concepts and analytical methods that can be used to estimate the employment effects of various energy choices. It also presents a general framework for determining how much information on employment effects is relevant to a particular policy issue in a given economic context, as well as detailed references for where the information is available and at what cost.

159
U.S. Department of Labor, Bureau of Labor Statistics
    1975   Impact of Federal Pollution Control and Abatement Expenditures
           on Manpower Requirements.   Bulletin 1836.   Washington, D.C.:
           U.S. Government Printing Office.

           Using an input-output analysis, the Bureau of Labor Statistics estimates the employment created (both direct and indirect) in various sectors for each billion dollars of expenditures on pollution control. Combining this information with estimates derived from expenditures in various sectors for pollution control, the study arrives at estimates of the total direct and indirect employment resulting from a projected outlay of $4.7 billion in 1980. The job impact is estimated to be a net increase of 193,000 jobs in 1980 over what would have been the case if no pollution abatement policy had been in effect. It must be noted that the study does not take into account offsetting effects on employment of higher taxes or borrowing required to finance these expenditures.

160
U.S. Department of Labor, Manpower Administration
    1974 Secretary of Labor's Report on the Impact of Energy Shortages
          on Manpower Needs. Washington, D.C.: U.S. Government Printing
          Office.

          This report provides detailed current and projected (to
          the end of the decade of the 1970s) impacts of energy shortages
          on employment by region, by occupation, and by economic
          sector. These data are similarly disaggregated, but in less
          detail, by age and sex. The report indicates that substantial
          current (1974) unemployment and lay-offs were due to energy
          shortages, especially in the automobile and related industries,
          among semi-skilled workers, and in the northern Midwest
          region. The long-range impact of energy shortages on employ-
          ment is expected to decline and to become relatively minor as
          shortages in equipment (e.g., for coal mining) are alleviated
          and as R & D (e.g., on solar energy) progresses and allows for
          increased employment in other sectors of the economy. The
          report also contains current and recommended future actions to
          ease employment impacts of the transition to an energy-scarce
          future.

161
U.S. Environmental Protection Agency
    1977 1976 Fourth Quarter Report of the Economic Dislocation Early
          Warning System. Washington, D.C.: U.S. Environmental Protec-
          tion Agency.

          In response to possible adverse impacts of environmental
          regulations on employment, the EPA instituted the Economic
          Dislocation Early Warning System, which reports on the number
          and size of plant closings caused by regulations. The report
          suggests that both the number of plants closed and the number
          of jobs lost is small. It is estimated that only 98 plants
          closed due to environmental regulations from January 1971
          through December 1976. The job losses associated with these
          closings were estimated to be 19,580. The lost jobs are
          concentrated in the Midwest, with over 25 percent of job losses
          occurring in this region.

162
U.S. News and World Report
    1977 "Safe water or jobs? A classic confrontation." U.S. News and
          World Report (February 7): 47.

          Many studies of economic effects of pollution control
          suggest that while the overall impact on the economy may be
          minimal, some localized effects may be severe. This article

examines the controversy surrounding one such locality, the
Mahoning River Valley in Ohio, where numerous steel mills dump
untreated effluent into the river and air between Youngstown
and Warren. Many of these plants are old and marginal, and
would be unprofitable if required to implement pollution
control. Residents seem to prefer jobs to a clean environment,
and thus far EPA has granted temporary exemptions to pollution
standards.

163

United Nations Association of the United States of America, Economic
Policy Council, Energy and Jobs Panel
     1980 Energy and Employment: Issues and an Agenda for Research.
     New York: Economic Policy Council of United Nations Associa-
     tion of the United States of America.

          This paper provides a cursory review of energy-employment
     interactions and a review of research conducted in this area.
     The panel found that gaps exist in the research and suggest an
     agenda for filling this void. The general topics offered as
     high-priority research needs are: (1) the overall impact of
     energy availability on the general economy, (2) opportunities
     available for substituting labor and capital for energy in dif-
     ferent industries and sectors of the economy, (3) the effects
     of overall energy availability and substitution on employment,
     including industry, regional, occupational, and income distri-
     bution impacts, and (4) employment impacts of government
     policies to aid adjustments to new energy realities.

164

Walker, Nolan E., and E. Linn Draper
     1975 "The effects of electricity price increases on residential
     usage of three economic groups: A case study." In Texas
     Nuclear Power Policies, Vol. 5: Social-Demographic and
     Economic Effects, pp. 102-126. Center for Energy Studies,
     Study No. 1. Austin: University of Texas.

          Marginal frequency analysis based on a July 1974 survey of
     a random sample (N=60) of households in Austin, Texas, was
     performed to determine the impact of price increases on income
     groups, behavior and attitudes, and electricity consumption
     changes. Data were gathered by personal interview and from
     electricity consumption records (from a utility company) for
     the previous two years.
          It was found that the number of lower-income households
     increasing their energy use equalled the number decreasing
     their energy use. For middle-income households, the number
     decreasing was greater than the number increasing their energy
     use. For upper-income households, the number increasing ran
     well ahead of the number decreasing electricity consumption--
     suggesting that upper-income groups are the least influenced by
     price rises. Middle-income groups seem to show the greatest
     price elasticity.

165
Wallick, Franklin
    1972 The American Worker: An Endangered Species.   New York:
          Ballantine Books.

          Wallick, the editor of UAW's Washington Report, presents a
          popularized account of occupational health.   He reports the
          fight by unionists to gain the implementation of OSHA and the
          control of certain specific workplace hazards.   The book con-
          tains an extensive bibliography from the popular press and
          union publications on workplace hazards and workers' efforts to
          gain the right to a safe working environment.

166
Walter, Ingo
    1975 International Economics of Pollution.   New York: John Wiley and
          Sons.

          On pp. 129-134, Walter analyzes the possible effects that
          the existence of different national environmental control
          regulations may have on multinational corporate decision
          making.   He argues that lower operating costs in countries with
          lax environmental regulations may create pressures for firms to
          relocate away from countries with strong controls.   He also
          points out that plant location is a complex issue involving a
          multitude of variables.   Since environmental control expenses
          usually are minor, averaging only five percent of production
          expenses, environmental regulations are not likely to be sig-
          nificant in the decision-making process unless a bottleneck has
          been reached.   Additionally, he argues that political pressure
          deriving from loss of employment in the home country and
          questions regarding the social responsibility of exporting
          pollution problems will further impede relocations based on
          environmental regulation differentials.

167
Wasserman, Harvey
    1980 "800 Unionists give the lie to labor's rigid pro-nuke image."
          In These Times (October 22-28): 9, 22.

          This article is an account of a conference on labor
          opposition to nuclear power organized in part by the Environ-
          mentalists For Full Employment, a group trying to organize
          coalitions between unions and environmentalists.   Several
          instances of labor opposition to nuclear projects are reported,
          and the antinuclear position of several unions, including the
          UAW, UMW, and Machinists, is detailed.   The author is hopeful
          that once union members realize the long-term health risks,
          economic problems, and disemployment associated with nuclear
          power, unions will increase their liaisons with environmen-
          talists in an effort to stop nuclear power projects.

168
Watts, Glenn E.
    1980 "Energy & jobs."   Public Power 38 (July/August): 68-70.

The potential impact on jobs and workers of various approaches to attaining energy self-sufficiency is examined. While overall impacts are not expected to be significantly negative, changes in the distribution of jobs will cause substantial dislocations. Responsible planning to minimize the hardships to workers and their families occasioned by such adjustments is urged. Programs with these goals which have been implemented by western European nations are reviewed.

169
Wayburn, Peggy
    1978 "The redwoods: Jobs and the environment." Environment 20
          (April): 34-9.

          After 125 years of intensive logging operations, the old-growth redwood forest is almost cutover. The timber industry (which historically has been overtly antilabor) has formed a coalition with labor unions to block inclusion of some of the last old-growth forest into Redwood National Park. Industry and labor claim that continued cutting of old-growth forest is necessary to maintain employment levels in the area until replanted forest is mature enough for lumbering.
          Wayburn argues that unemployment in the area is high despite recent record timber harvests because the logging industry has become increasingly automated. Thus, employment will suffer even if logging continues. She infers that the profit motive may be at the heart of the industry's campaign to continue logging, noting that demand for the export market has driven up the price of redwood (and other) timber products in recent years.
          Expansion of Redwood National Park would, she argues, do more to stem unemployment than continued logging would. She cites the obvious employment opportunities from tourism as well as a plan to rehabilitate logged-over areas which would require traditional logging skills. She concludes that in this case, as in many others, the issue of jobs vs. environment is more illusory than real.

170
Wilson, Margaret B.
    1978 "Energy growth and jobs." Paper presented at Edison Electric
          Institution 46th Annual Convention, Houston, April 10-12.

          The author presents the National Association for the Advancement of Colored People's (NAACP) position on Carter's National Energy Plan. The plan would restrict economic and energy growth which would, in turn, increase the already high unemployment rates among the poor and blacks. The NAACP supports instead a policy of aggressive economic expansion which is in the mutual interest of the energy industry and the NAACP.

171
Winger, Jon, and Carolyn Nielsen
    1976 Energy, the Economy and Jobs. Special Report from Chase
          Econometrics, New York, September.

The energy problem is presented as a political problem
resulting from incomplete exploitation of domestic energy
sources. Complacency on the part of both consumers and the
government presents a barrier to solving this problem.   The
relationship of energy consumption to GNP is developed to
illustrate the effects of current shortages on employment and
the economy in general.

172
Winter, Lester A.
   1977 "Balancing act: Minnesota project shows jobs vs. ecology
          dilemma." Wall Street Journal, November 15: 1, 20.

          1973 court rulings forced Minnesota iron mines to stop
          dumping waste rock in Lake Superior based on evidence that the
          fine, asbestos-like particles in the waste may cause cancer. A
          50 percent rise in the world price of iron pellets finally made
          feasible a project to eliminate the waste from the water and
          dust in the air.   Construction of the $370-million project,
          costing more than the plant itself, ends a six-year battle
          between environmentalists (allied with government officials)
          and local employees concerned about possible loss of jobs.   In
          fact, the pollution control project will employ an additional
          65 workers, not including the employment generated during con-
          struction of the facility.

173
Woodcock, Leonard
   1972 "Labor and the economic impact of environmental control
          requirements."   In Jobs and the Environment: Three Papers.
          Berkeley: Institute of Industrial Relations, University of
          California.

          Woodcock, president of the United Auto Workers, argues
          that stringent environmental regulations and enforcement are
          necessary. Woodcock is critical of the Nixon administration
          for espousing environmental control but colluding with industry
          in sidestepping environmental regulation.   He endorses the
          principle that ". . . the burdens and sacrifices required by an
          action taken in the service of the interests of the whole
          society should be shared equitably by all who benefit from that
          action and not be allowed to fall disproportionately on some,
          who are made victims of the action" (p. 10).
          Woodcock feels that environmental and social change is
          made difficult by large, socially indifferent, and intransigent
          corporations.   He calls for consistent public control through
          governmental review of industry plans in terms of employment
          and environmental and social impacts. Woodcock notes that this
          is already the case in many European countries. Woodcock feels
          that trade unions can act through collective bargaining to
          achieve the above goals. However, there will be a limit to the
          effectiveness of collective bargaining in achieving environ-
          mental goals; therefore, government legislation is necessary.

174
Zorzoli, G. B.
    1977 "Some considerations on the opportunities offered to R & D in
          the field of labor intensive technologies." Energia Nucleare
          24 (January): 35-37.

          The author discusses the feasibility of labor-intensive
          technologies in advanced industrial societies. His conclusion
          is that such an approach is possible and that it offers several
          advantages over conventional capital-intensive production.
          Among these advantages are the energy savings, the ability to
          transfer human labor between diverse production tasks, the
          lower transportation costs associated with the decentralized
          nature of labor-intensive systems, and the lower social welfare
          needs of such a system.

Part II

Benefit-Cost Analysis:
A Labor-Environmental Concern

CHARLES C. GEISLER

The American workplace is a proving ground for new technologies. The consequences of these technologies vary widely--some good, some bad, many unknown. Today, a social-environmental contract exists to protect people who mingle with these new technologies on and off the job. The contract consists of a legislative bulwark developed during the past two decades to champion broadly defined rights of personal and environmental health. But the bulwark is corroding from massive deregulatory initiatives within the federal government. Central to the deregulatory effort is a controversial decision-making tool known as benefit-cost analysis (BCA).

This essay offers a brief introduction to BCA, its history, its appeal amid economic belt tightening, and its many frailties. It suggests that BCA has already undergone careful scrutiny by environmentalists, a scrutiny from which labor interests can benefit. At the same time, it cautions against mechanical indictments of BCA sometimes leveled by its critics. The objective of BCA is not always deregulation, despite its recent use by fiscal conservatives to overturn legislation pertinent to the environment (e.g., EPA v. National Crushed Stone Association) and to occupational health and safety (e.g., American Textile Manufacturers' Institute v. Marshall and National Cotton Council v. Marshall).

Operationally, BCA entails several steps: identification of policy effects, their classification into benefits and costs, their quantification in a common measure such as dollars, the discounting of the future effects into present terms, and, finally, the weighing of costs and benefits to see which is greater (U.S. Congress 1980). This technical description carries with it an air of objectivity and rationality. Yet if BCA experience in nonworkplace environments serves as a guide, the potential for BCA misuse is broad indeed. Under the best of circumstances, BCA can greatly oversimplify complex social and environmental realities. It has been used self-servingly by government agencies to promote certain public works and programs and to demote others. The public interest is juggled and jostled in the process. "Costs" and "benefits" are balanced as much by political powerbrokers as by enlightened public servants.

Like the causal connections between industrial pollution and environmental degradation, cause and effect between workplace conditions and deteriorating worker health are difficult to establish. Yet these effects are not fictitious. On-the-job accidents kill an estimated 13,000 workers annually, and 9 million additional workers either require medical care or are temporarily unable to work because of work-related injuries. These injuries translate into 245 million workdays lost each year—ten times more than the average number of days lost annually through strikes (Congressional Quarterly, 1981). In 1979, the National Safety Council placed a total national per-annum cost on work-related injuries and fatalities in 1979 at some $30 billion.

BCA: The Birth of a Notion

The latter part of the nineteenth century and the early decades of the twentieth century were the incubation phase of benefit-cost analysis. As public works consumed ever greater portions of the federal budget, a decision-making tool was needed to select among public-works alternatives in the absence of a private-market mechanism to make these choices. BCA's expanded use since the 1930s, when a new wave of public-works expenditures became integral to the national economy, has largely been due to its popularity among economists and engineers. Gradually, BCA came into general use as a decision tool for selecting public services. But a basic change was in the offing. When enlisted to evaluate public works such as roads, harbors, or dams, BCA was used (as environmentalists would discover) by behemoth government agencies to endorse their expansionary plans. As a diagnostic device for regulations protecting consumers, labor, or the environment, BCA succumbed to far more conservative, nonexpansionary tendencies.

The Occupational Safety and Health Act of 1970, covering roughly 5 million employers and 60 million employees in a broad range of occupations, introduced a new principle into American life—that workers have the right under federal law to a safe place to work (Congressional Quarterly, 1981). In 1974, President Ford set the stage for an eventual BCA-OSHA confrontation through Executive Order 11821. This required all major regulatory legislation to be subjected to "Inflation Impact Statements" by the Office of Management and Budget and by the Council on Wage and Price Stability. Regulations, including those of OSHA, were taken to be inflationary if their costs exceeded their measured benefits. When President Ford left office, complete OSHA standards had been issued to control only seventeen of the thousands of workplace substances menacing workers' health.

President Carter, also plagued by an inflationary economy, continued the policies of his predecessor. By and large, regulations were justified only if cost-effective. Qualified administrators were appointed to run OSHA (Carter chose a trained toxicologist) and to cut red tape. Using epidemiological studies, Carter appointees developed new health standards to protect cotton-mill workers from brown lung disease and workers in other industries from the effects of such substances as benzene, arsenic, and lead. BCA, while seen as valuable, was not the ultimate measure in eliminating or retaining regulations. Charles Schultze, Chairman of Carter's Council of Economic Advisors, told the Senate Government Affairs Committee in 1979: "The evaluation of costs and benefits in regulatory analysis should not be a simple-minded attempt to measure benefits in dollar terms, to measure the costs, and then make some mechanical decision on the basis of comparing the two sums." (Ruttenberg 1981).

The 1970s, frequently depicted as "The Environmental Decade," were also years of national economic trauma in which wages and profits, rising together since World War II, would cease their cordial relationship. A climate of recession loomed large in most economic forecasts by

1974, rousing strong opposition to environmental regulations by the business community. Where regulation appeared to stifle jobs, as business insisted it did, labor shared this opposition. Though many industries prospered by manufacturing pollution-monitoring and abatement equipment and though there was ample evidence that regulatory costs were exaggerated, many Americans were convinced by the late 1970s that they must choose between economic and environmental health (Kazis and Grossman 1982). Many construed the Carter defeat as a repudiation of the social-environmental contract so long in the making.

With the Reagan election victory, business lobbyists in Congress and virtually all of Mr. Reagan's cabinet appointees redoubled efforts to apply BCA to a broad range of federal regulation. Two days after being sworn into office, Reagan announced a new Presidential Task Force on Regulatory Relief consisting of his closest advisors. The following month, the president signed Executive Order 12291 providing that "regulatory action shall not be taken unless the potential benefits to society outweigh the potential costs to society." He justified the order before a joint session of Congress, claiming, among other reasons, that regulation fosters high unemployment. Specifically, the order required four things:

--that executive agencies prepare regulatory impact analyses of potential costs and benefits for all new rules which could lead to "major increases" in costs to consumers or industry, and that they also describe lower-cost alternatives;

--that all "major" new rules already proposed be postponed until the above analyses were completed, except in cases where statutory or court-imposed deadlines or emergencies existed;

--that the Office of Management and Budget was authorized to designate regulations as "major" and to require the agencies to consider additional evidence and information in making their decisions; and

--that agencies must identify for review, change, or recision any existing rules that did not follow the least costly approach.

As one observer noted, such requirements might ameliorate concerns of some regulatory reformers but would not solve--and might exacerbate--concerns that were arguably more fundamental (Andrews 1981).

The Supreme Court on BCA

Encouraged by shifting views on public regulation, business interests acted quickly to test the legality of standards protecting both the workplace and the larger environment--such regulations, according to business, infringed on property rights or unduly limited states' powers (Congressional Quarterly, 1981). Specifically, the textile industry opposed OSHA's cotton-dust exposure standards and the National Crushed Stone Association opposed the clean water standards imposed by EPA. Litigation ensued with profound consequences for both. In December 1980, the Supreme Court unanimously held that EPA was not bound by economic considerations of water-pollution law enforcement. Six months later, a five-to-three majority of the Court upheld OSHA's

cotton-dust standard over industry contentions that the standard was
invalid because OSHA failed to apply BCA during promulgation (Commerce
Clearing House 1981).   The ruling was heralded by organized labor as one
of its principal legal victories.   Justice Brennan wrote in the decision
that the health of workers outweighs "all other considerations" and that
"any standard based on a balancing of costs and benefits would be
inconsistent with the law."

The Reagan administration's noticeable support for BCA as a check
on regulation must be put in the context of a reindustrialization dream
taunted by inflation.   The executive order cited earlier sought to
reduce costs to business and thereby curb inflation.   The Chairman of
the Council of Economic Advisors had informed President Reagan that
regulation was costing the nation more than $100 billion per year (else-
where disputed; see Kazis and Grossman 1982).   Thus, few were surprised
to find that the new administration promulgated only thirty regulations
during its first year in office.   This contrasts to the 100 to 200 such
regulations customarily enacted each year by presidents of either
political party (Shabecoff 1981).

Neither OSHA nor EPA, as the Supreme Court affirmed, are legally
bound by BCA.   In establishing OSHA, Congress committed the country to
strict safety and health standards in the workplace, with stiff penal-
ties for employers who fail to comply.   In essence, OSHA represented a
commitment to an emerging worker's right: no worker should have to face
injury or illness in order to make a living.   Yet the Supreme Court's
verdict did not preclude BCA.   The Court's ruling left the discretionary
use of BCA an unsettled question. While not mandatory, BCA might be
permitted (Deland 1981).   Moreover, the Court held that workers' health
must be safeguarded "to the extent feasible," which implies economic
considerations. And in practice, the challenging task of enforcing OSHA
is delegated to state-level OSHA offices where enforcement is apt to be
somewhat discretionary (Crawford 1981).   Finally, though the Supreme
Court has acted to preserve the original intent of OSHA and the Clean
Water Act, at least three of the justices are in their seventies and
nearing retirement, and Congress may amend the laws in question.

The class implications of Mr. Reagan's actions leave little to
conjecture.   In 1979, the chief executives of nearly 200 major companies
known collectively as the Business Roundtable released a detailed study
indicting the regulatory costs of selected federal agencies, including
EPA and OSHA (Kazis and Grossman 1982).   Shortly after Mr. Reagan's
election, his designated Labor Secretary withdrew a major rule requiring
the labeling of hazardous chemicals in the workplace, and the administration expressed its intention to liberalize the Department of
Interior's surface-mining regulations and amend the Clean Air and Water
Acts.   Mr. Reagan's appointed OSHA Director exempted the construction
industry from requirements that it provide medical records to its
workers; delayed requirements that the smelting industry comply with
reduced levels of lead in employee blood; drastically curtailed worker-
educational programs and materials such as an OSHA booklet on brown lung
disease written for textile workers; and put a halt to many health
standards being developed by OSHA (Smith 1981).   The same OSHA director
acknowledged the consulting assistance of Organization Resource
Counselors, Inc., an industrial-relations firm supported by sixty major
corporations and the source of vociferous complaints against OSHA
regulations on lead, noise, and worker medical records.

BCA Problems: Temporary and Long-Term

Despite the overwhelmingly probusiness stance of the Reagan administration with respect to regulation, it would be shortsighted to conclude that BCA has an inherent conservative or even class bias. It is often difficult to distinguish philosophically between BCA's opponents and proponents except in specific historical circumstances. Quite clearly, Mr. Reagan's deregulation efforts using BCA have induced marked centralization of authority in the executive branch—a tendency not all conservatives will abide. Nor is BCA philosophically rooted in Republican ideals. It originates in a tradition of public-investment economics requiring progressively more governmental regulation rather than less (Andrews 1982). It is not unthinkable, therefore, that Republicans will continue to embrace deregulation but disown BCA in the future. Moreover, in years to come, BCA may be employed by environmentalists and labor representatives to defend their respective and joint interests. With this in mind, the remainder of this essay addresses common criticisms to which BCA, as the following annotations make clear, has been subjected.

First among these criticisms is that BCA is deceptively precise. Because it is quantitative, its practitioners and the public are inclined to accept its designation of what is a "cost" and what is a "benefit" without misgivings. Yet each of the steps in the operational definition offered at the beginning of this essay has technical pitfalls. Each is subject to questionable manipulation when practitioners seek, as BCA requires, to convert the quality of life to the quantity of life. Even proponents with the best of intentions can (and do) err in assigning dollar values to such things as morbidity and mortality in the workplace or among endangered species. Several years ago Congressman Paul Rogers (D-Fla.) warned against uncritical application of strict BCA to worker-health policies. The result, he cautioned, could be a decision that "if getting sick is cheaper, then maybe we should not try to prevent illness." (Congressional Quarterly, 1981:4). Needless to say, many developers who consider a sick environment an acceptable cost of progress and profits would unashamedly agree.

This narrowing of social reality to accommodate economic efficiency criteria produces administrative rigidity, another criticism of BCA. There is a tendency in BCA to relegate less tangible social costs to "externality" or unmeasurable categories, since their very measurement is prone to add inefficiency and delay to the administrative process. Matters such as worker alienation in exchange for more efficient assembly-line production, or workers' rights to medical records in exchange for reduced operating costs and consumer prices, are difficult to quantify in other than conventional terms of profit or loss. The net effect of such convention is to narrow public debate and consciousness, to sanctify efficiency-producing technologies, and to keep BCA the domain of "experts." Mechanisms for public involvement in BCA are neither fostered nor missed.

This second criticism invites a third: BCA for whom? BCA can only conscientiously serve the public interest to the extent that its logic and assumptions are broadly held public property. Realistically, its application often reflects the balance of political power in a particular setting; in the mindful eyes of one skeptic, BCA is about as neutral as a voter literacy test in the Old South (Congressional Quarterly,

1981). If the government and its most influential constituents desire a program (say, the condemnation of working-class homes to provide growing space for major industry, as in the Poletown area of Detroit), BCA can be manipulated to affirm that benefits exceed costs. If, to cut costs, the same government wishes to terminate regulations protecting workers from toxic substances, BCA is likely to show costs in excess of bene-fits. In this sense, the scientific neutrality of BCA can be mislead-ing. What should be included as benefits and costs, what weights should be assigned to each, who will conduct the analysis, and what discount rates should be selected--all these questions are apt to have answers which are at least in part political. BCA will, by extension, serve the public interest best in situations where power is genuinely shared.

The final ratio of benefits to costs hinges on the discount rate--a fourth source of criticism and controversy surrounding BCA. The discount rate--a technical device for converting future benefits of a project or program to their present equivalents--is a means used by economists to compare value of policy alternatives over time. In recent years there has been a trend in government to make discount rates "official"--that is, to base them on the nominal cost of federal funds as indicated by the yields of U.S. bonds. The greater the inflation rate and the more sustained the inflation, however, the more inflated is the official discount rate. Since, as a rule of thumb, the higher the rate of discount, the more "discounted" (i.e., reduced) is the value of something tomorrow compared with the same item today, inflation in the U.S. economy automatically reduces the average annual benefits and raises the average annual costs of public investments (Mugler 1982). There is little guarantee, as certain of the following works indicate, that unofficial discount rates would greatly improve this state of affairs.

This list of BCA pitfalls and imperfections is not exhaustive. Nor does it emphasize sufficiently that labor must not satisfy itself with existing criticisms of BCA. The literature which is annotated below frames many BCA issues, including the difficult tasks of placing values on worker and citizen rights and of selecting judicious discount rates. It transmits viewpoints implacably opposed to BCA, yet also offers new approaches to quantifying indirect costs and benefits--approaches intended to reform BCA. Other literature cited challenges certain cost-accounting schemes used by industry, on the one hand, and suggests cost savings awaiting industries that eliminate health and safety hazards on the other. When OSHA moved in 1975 to regulate the carcinogen vinyl chloride, for example, a study commissioned by plastics manufacturers predicted that compliance would cost $90 billion. This estimate turned out to be 300 times too high, and many companies producing vinyl chloride made handsome profits after the regulation was enforced (Crawford 1981). Ruttenberg and Hudgins (1981) documented a substantial drop in the rate of recorded injuries and illnesses during the 1970s among workers in the chemical industry, the most hazardous in the country, and attribute the decline to OSHA inspections. Many more examples of BCA-related cost savings appear in Kazis and Grossman (1982).

Conclusion

As expressed earlier, there is a widespread impression held over
from the environmental decade of the 1970s that the economy and the
environment, broadly defined, are hopelessly at odds. Some would argue,
as have Mr. Reagan and many of his strategists, that the same can be
said of economic and social programs, including regulation of the workplace environment. By such logic, efficiency and equity are mutually
repellent. The literature annotated in this volume challenges these
homilies. Contrary to impressions that social—environmental and
economic objectives are antagonistic, there are numerous cases where
greater social equity produces greater economic efficiency rather than
less—e.g., worker-owned industries. Healthier and safer workplaces
should be added to this list.

BCA flourished in the Great Depression and will undoubtedly be a
hallmark of the recession now at hand. The Reagan years will be remem-
bered by labor and environmentalists alike as a time of severe regulatory challenge in the name of regulatory reform. Many regulations,
attained only after years of public organizing, bill drafting, lobbying,
and sacrificed individual and environmental health, are being ridiculed
and emasculated without public consultation. Like so much of the
vocabulary of government, reform has come to mean its opposite.

These observations need not demoralize labor's attempts to counter
antiregulatory uses of BCA. The raw political fact is that American
working people will continue to constitute a major political force
through most of this century, and are powerfully positioned to elect a
series of forthcoming presidents and thousands of future congressional
representatives (Levinson 1976). Labor's alliance with environmentalists, a movement with a new lease on life, bodes well for the forseeable
future. This being so, labor will surely be in a position to shape the
direction and quality of BCA. Given the probability that the workplace
will continue to be an arena of BCA contention in the years ahead, the
annotations which follow should be of both immediate and long-term use.

References

AFL-CIO
    1980   "Fact sheet on the Schweiker-Williams 'Occupational Safety and Health Improvement Act of 1980.'" Committee to Save OSHA, P.O. Box 1666, Grand Central Station, New York.

Andrews, Richard N. L.
    1981   "Will benefit-cost analysis reform regulations?" Environmental Science and Technology 15 (September): 1016-21.

    1982   "Cost-benefit analysis as regulatory reform." In Cost-Benefit Analysis and Environmental Regulations: Politics, Ethics and Methods, ed. Daniel Swartzman, Richard A. Liroff, and Kevin G. Croke, pp. 107-36. Washington, D.C.: The Conservation Foundation.

Commerce Clearing House
    1981   "Supreme Court upholds cotton dust standard, rejects cost-benefit test." Employment Safety and Health Guide, No. 258 (June 23):1-3.

Congressional Quarterly, Inc.
    1981   Environment and Health. Washington, D.C.: Congressional Quarterly, Inc.

Crawford, James
    1981   "Court saves health standard--for now." In These Times (July 15-28):5.

Culyer, Anthony J.
    1978   "The quality of life and the limits of cost-benefit analysis." In Public Economics and the Quality of Life, ed. Lowdon Wingo and Alan Evans, pp. 141-53. Baltimore: Johns Hopkins University Press.

Deland, Michael R.
    1981   "Cost-benefit analysis and environmental regulations." Environmental Science and Technology 15 (September):997-1002.

Kazis, Richard, and Richard L. Grossman
    1982   Fear at Work. New York: The Pilgrim Press.

Levinson, Andrew
    1976   "The working-class majority." In Crisis in American Institutions, ed. Jerome H. Skolnick and Elliott Currie, pp. 139-55. Boston: Little, Brown and Company.

Mugler, Mark W.
    1982   Effects of the Discount Rate on the Civil Works Program. Policy Study 82-0900. Fort Belvoir, Va.: U.S. Army Engineer Institute for Water Resources, July.

Ruttenberg, Ruth
    1981   "Why social regulatory policy requires new definitions and
            techniques for assessing costs and benefits: The case of
            occupational safety and health." Labor Studies Journal 6
            (Spring):114-23.

Ruttenburg, Ruth, and Randall Hudgins
    1981   Occupational Health and Safety in the Chemical Industry.
            New York: Council on Economic Priorities.

Shabecoff, Philip
    1981   "Reagan order on cost-benefit analysis stirs economic and
            political debate." New York Times, November 7:28.

Smith, R. Jeffrey
    1981   "OSHA shifts direction on health standards." Science 212
            (June 26):1482-83.

U.S. Congress, House Committee on Interstate and Foreign Commerce
    1980   Use of Cost-Benefit Analysis by Regulatory Agencies. Joint
            Hearings, Subcommittees on Oversight and Investigations and
            on Consumer Protection and Finance. Ninety-Sixth Congress,
            First Session (July and October, 1979). Serial No. 96-157,
            pp. 11-12. Washington, D.C.: U.S. Government Printing
            Office.

Bibliography II:
Benefit-Cost Analysis and
Environmental Initiatives

compiled by
IRVING W. WISWALL and CHARLES C. GEISLER

175
Andrews, Richard N. L.
 1981 "Will benefit-cost analysis reform regulations?" Environmental Science and Technology 15 (September): 1016-1021.

 BCA is currently being offered as a means of reforming the regulatory process, especially among those agencies that protect health, safety, and environmental quality. Andrews expresses doubt that the inclusion of BCA will aid in meeting its advocates' objectives, and suspects that its use will substantially detract from the purpose of the regulation. At its best, BCA is capable of great comprehensiveness and precision. However, in general application, it is an inexact and manipulable method of analysis. It is subject to arbitrariness, subjectivity, and self-interest. As such, it can be expected to reflect political and institutional dynamics rather than a rational accounting of costs and benefits.

 In light of his analysis, Andrews expects the following: (1) the "worst" regulations, in terms of cost ineffectiveness, will be stopped, although numerous others with desirable consequences will also be hampered; (2) some regulations will be acted on in a more cost-effective manner, if not on the basis of formal B/C analysis; (3) new regulations will be retarded; and (4) BCA will further centralize power in the presidency and in the Office of Management and Budget.

176
Arrow, Kenneth, and Anthony C. Fisher
 1974 "Environmental preservation, uncertainty and irreversibility." Quarterly Journal of Economics 88 (May): 312-319

 This paper addresses the deterministic treatment of processes that are in fact probabalistic. The authors show that, 1) if development involves irreversible transformations of the environment resulting in a loss in perpetuity of benefits flowing from a preserved environment, and 2) if information about costs and benefits of alternatives is imperfect, the projected

benefits from development will generally be inflated. They conclude that benefit calculations of an irreversible decision made
with imperfect information should be adjusted downward to
reflect the loss of options entailed.

177
Bailey, Martin J.
    1980 Reducing Risks to Life: Measurement of the Benefits. Washing-
    ton, D.C.:   American Enterprise Institute for Public Policy
    Research.

        The argument here, based on a review of five separate
    studies, is that existing government programs for health and
    safety will save more lives if officials use BCA to guide pro-
    gram decisions. Presently, some regulatory agencies neglect BCA
    and others lack uniform BCA methods, greatly weakening their
    agency effectiveness; BCA problem areas are acknowledged. The
    Office of Management and Budget should enforce high standards of
    BCA quality and reliability. Regulation without BCA, paradoxi-
    cally, threatens to reduce economic efficiency, our standard of
    living, and ultimately our ability to save lives from accidents
    or ill health.   That is, failure to regulate regulators--by
    imposing BCA guidelines--leads to costly solutions saving a few
    lives compared to much broader reductions in morbidity and
    mortality which would follow from less regulation aimed at a
    broader spectrum of health hazards. In sum, a limited amount of
    regulation is more socially beneficial (less costly) than an
    unlimited amount--the trend until very recently. The author is
    a professor of economics at the University of Maryland, and
    integrates both social and environmental material in the course
    of his argument.

178
Ball, Michael
    1979 "Cost-benefit analysis: A critique." In Issues in Political
    Economy, ed. Francis Green and Peter Nore, pp. 63-87. London:
    Macmillan Press.

        BCA is based on certain aspects of neoclassical economic
    theory.   The implementation of that theory as a basis for
    measuring societal preferences is based on a series of assump-
    tions which Ball asserts are arbitrary and politically and
    ideologically, not rationally, based.   Ball presents these
    assumptions and points out their internal inconsistencies and
    poor fit with "real world" behavior. His conclusion is that
    these assumptions are adopted simply to enable a neoclassical
    model of market economy to be used. Ball's analysis is unusual
    in that he does not rest his case with the demonstration of the
    shortcomings of B/C analysis. His second task is to explicate
    the adoption of this and similar methodologies by the state
    apparatus of advanced capitalistic countries. He develops the
    idea that the purpose of B/C analysis is to aid the state in its
    goal of maintaining conditions for the accumulation of capital
    by dominant elites.   A further implication of the use of B/C
    analysis is the diversion of class struggle away from real
    issues to superficial questions.

179
Baram, Michael S.
    1980 "Cost-benefit analysis: An inadequate basis for health, safety,
          and environmental decision making."   Ecology Law Quarterly
          8: 473-531.

          This article reviews the methodological limitations of
          B/C analysis, its current agency use, and the impact of recent
          executive orders mandating B/C analysis of all major regulatory
          decisions. The author concludes that if B/C analysis is to
          provide a reasonable foundation for decision making, greater
          public accountability and participation are needed.

180
Baram, Michael S.
    1976 "Regulation of environmental carcinogens: Why cost-benefit
          analysis may be harmful to your health." Technology Review 78
          (July/August): 40-42.

          Several regulatory agencies are responsible for controlling
          environmental carcinogens. Each justifies the release of some
          carcinogenic matter into the environment on the basis of B/C
          analysis. The cumulative effect of each small contribution may
          well be substantial.   A more holistic approach by the
          legislature is needed to overcome this problem.

181
Bator, Francis M.
    1957 "The simple analytics of welfare maximization."   American
          Economic Review 47 (March): 22-59.

          This article is a clear, concise, nonmathematical summary
          of the basic principles of welfare economics. Since this branch
          of economics provides the theoretical underpinnings of B/C
          analysis, it may be of interest to those seeking a more detailed
          understanding of B/C analysis.

182
Baumol, William J.
    1968 "On the social rate of discount."   American Economic Review
          58:4 (September): 788-802.

          The basic model Baumol employs in his discussion is that
          the discount rate should be chosen such that social benefits
          from public spending exceed the forgone benefits had the funds
          remained in the private sector. Within this framework, a good
          discussion of the factors that should be considered when
          choosing a discount rate is presented. For comments on this
          approach see: American Economic Review 59:5: Alan Nichols,
          pp. 909-911; Estelle James, pp. 912-916; Carl Lanauer, pp.
          917-918; David Ramsey, pp. 919-924; Dan Usher, pp. 925-929.
          Baumol responds to these comments briefly on p. 930.

183

Bennington, John, and Paul Akelton

   1975   "Public participation in decision-making by governments."   In
          Benefit-Cost and Policy Analysis 74, ed. Richard Zeckhauser,
          Arnold A. Harberger, Robert H. Haveman, Laurence E. Lynn, Jr.,
          William Niskanen, and Alan Williams, pp. 417-455.   Chicago:
          Aldine Publishing Company.

          Formal modes of evaluation, like B/C analysis, exclude the
          working class from the decision-making process. If analysis is
          to live up to its claim of including individual preferences,
          methods must be developed to involve affected individuals in the
          process.   The usual avenues for citizen participation do not,
          the authors argue, meet this need. Current modes of participa-
          tion pacify citizens rather than elicit information.

184

Bohm, Peter, and Claude Henry

   1979   "Cost-benefit analysis and environmental effects."   Ambio 8:
          18-24.

          B/C analysis can be used to reduce the dimensions of a
          problem and it can also avoid the common alternative decision-
          making problem of unidimensionality. The authors urge inclusion
          of distributive effects and so-called option value in B/C
          analysis techniques designed to deal with risk and uncertainty,
          since avoiding a decision with irreversible consequences is in
          itself a value.   The article concludes with three case studies,
          one in which B/C analysis was not but could have been employed,
          and two in which the analysis was profitably employed.

185

Broadway, Robin W.

   1974   "The welfare foundations of cost-benefit analysis."   Economic
          Journal 84 (December):   926-939.

          Advocates of BCA assume that if the results of an analysis
          show that the benefits outweigh the costs of a proposed action,
          net societal welfare will increase. The assumption behind this
          belief is that net gainers would compensate net losers and the
          change would result in a net improvement. The purpose of the
          paper is to show that, even if this compensation criterion is
          accepted as appropriate, the simple summation of money gains and
          losses does not satisfy this test.   Broadway's conclusion is
          based on the observation that the marginal utility of income is
          not equal for all income groups. His suggested solution is to
          incorporate distributional weights into the summation of costs
          and benefits, based on estimates of the marginal utility of the
          effects.

186

Bromley, Daniel W., and Bruce R. Beattie

   1974   "On the incongruity of program objectives and project evalua-
          tion: An example from the reclamation program."   In Benefit-
          Cost Policy Analysis 73, ed. Robert H. Haveman, Arnold C.
          Harberger, Laurence E. Lynn, Jr., William A. Niskanen, Ralph

Rurvey, Richard Zeckhauser, and Daniel Wisecarver, pp. 144-148. Chicago: Aldine Publishing Company.

The authors contend that project evaluation is often based on narrower criteria than overall objectives of the program. Efforts to develop measures of nonmonetary benefits should continue in order that program objectives not strictly related to market efficiency have an ongoing place in projects and project analysis.

187
Cicchetti, Charles J., Robert K. Davis, Steve H. Hanke, and Robert H. Haveman
   1973 "Evaluating federal water projects: A critique of proposed standards." Science 181 (24 August): 723-728.

The authors' principal objections to proposed guidelines for evaluating federal water resource projects are the guidelines' implicit misstatement of certain fundamental economic principles and recommendations of faulty estimation procedures. The authors find the procedures for benefit estimation lacking; for example, certain opportunity costs are virtually ignored, especially those related to the environment. They note the unreasonably low choice of discount rates as well. Judgments regarding environmental effects, regional development effects, and equity effects are left to agency personnel without explicit guidelines. Finally, the authors note that in ignoring cost sharing and pricing policies, the guidelines are likely to encourage inefficient and inequitable projects. They believe the cumulative effect of these deficiencies is a bias toward excessively large projects. Their analysis is indirectly relevant to B/C logic.

188
Clark, Elizabeth M., and Andrew VanHorn
   1976 Risk-Benefit Analysis and Public Policy: A Bibliography. Updated and extended 1978 by Laura Hedal and Edmund A. C. Crouch, Informal Report # BNL 22285-r. Upton, N.Y.: Brookhaven National Laboratory.

This extensive (99-page) bibliography contains both theoretical and applied works on risk-benefit analysis. The methodology can be viewed as an extension of B/C analysis, of use when risks and benefits are difficult to evaluate in dollars. Several articles which adopt a critical stance toward risk-benefit analysis are listed separately in the present work.

189
Congressional Quarterly, Inc.
   1981 Environment and Health.   Washington, D.C.: Congressional Quarterly, Inc.

This well-documented volume summarizes the major environmental policy issues of the 1980s. It gives an overview

of the growth in environmental awareness that occurred in the
'70s, and of the reduction of its strength with the advent of
the Reagan administration and a new Congress. Air, the land,
water, nuclear power, food, and the workplace are examined as
separate environmental issues. The concluding chapter evaluates
existing regulation and attempts at deregulation. The
appendixes include a chronology of environmental legislation
from 1970 to 1980; selected documents, including court
decisions, agency reports, and scientific studies; and a
bibliography.

Cost-benefit analysis is referred to throughout the book.
As observed in the introduction, "Probably the most emotional
aspect of the environmental movement arises from the difficulty
of placing a dollar figure on the worth of human well-being, or
even a human life." Nonetheless, BCA and its recent relative,
risk-benefit analysis, are described as priorities for many
regulatory reformers, with the main source of political support
coming from the business community. Environmental groups see
BCA as "a smokescreen for cutting the heart out" of environmental protection programs.

BCA is covered in greatest detail in chapter 7, on the
environmental dangers in the workplace--the area of many court
challenges to regulation based on the lack of BCA reasoning. In
the concluding chapter, BCA is examined further, and both sides
of the debate are summarized.

190
Culyer, Anthony J.
  1977 "The quality of life and the limits of cost-benefit analysis."
         In Public Economics and the Quality of Life, ed. Lowdon Wingo
         and Alan Evans, pp. 141-153.   Baltimore: Johns Hopkins
         University Press.

Culyer argues that there are logical limits to the
applicability of BCA in the design of social policy. A
principal objective of social policy is the creation of
institutions that foster integration and discourage alienation
in society. This goal may well conflict with the most broadly
defined efficiency objective. Given these conflicts, the design
of social policy must often make trade-offs between arrangements
that are just and those that are efficient. BCA is useful for
arriving at quantification of efficiency but it cannot guide
decision making involving trade-offs between justice and
efficiency. BCA's second limitation derives from theoretical
inability, in the absence of a market, to make interpersonal
comparisons of utility with sufficient accuracy to provide
useful policy guidelines.   Attempts to incorporate equity
considerations and interpersonal utility measures into BCA are
logically inconsistent and, when practiced, result in confusion.
Worse, such attempts lend the weight of scientific authority to
purely personal prejudices. Culyer suggests that economists
limit their BCA strictly to market efficiency and leave equity
and interpersonal utility matters to noneconomists.

191
De Wet, G. L.
   1976 "The state of cost-benefit analysis in economic theory." <u>South
   African Journal of Economics</u> 44 (March):   50-64.

         This article attempts to discern if B/C analysis is rooted
   in widely accepted principles of economics.   The conclusion
   reached is that the technique is an excellent method of
   measuring advantageous and disadvantageous effects of economic
   action, but that it fails in pricing these effects.   De Wet
   concludes that B/C analysis is not based on sound economic
   theory.

192
Deland, Michael R.
   1981 "Cost-benefit   analysis   and   environmental   regulations."
   <u>Environmental Science and Technology</u> 15 (September):   997.

         Deland reviews the potential impact on environmental
   regulations of President Reagan's executive order that BCA guide
   regulatory action.   He notes that toxic materials legislation,
   the Clean Air Act, and the Clean Water Act all appear to
   prohibit the balancing of costs and benefits in the setting of
   standards.   Each act specifies that the best available
   technology be implemented to protect workers and the
   environment.
         In recent litigation challenging these laws, the Supreme
   Court and federal appellate courts have concluded that neither
   OSHA nor EPA are <u>required</u> to balance costs and benefits.   Court
   opinions have been unclear as to whether the legislation would
   <u>allow</u> the balance of costs and benefits, but they seem to imply
   that the legislation would not.   While this suggests that
   current regulatory acts will not be affected by the executive
   order mandating BCA, the order has given Congress a clear
   message on the administration's view of regulation.   The degree
   to which Congress heeds this message will shape future
   legislation.

193
Doeleman, J. A.
   1980 "On the social rate of discount: The case for macro
   environmental policy." <u>Environmental Ethics</u> 2 (Spring):   45-58.

         The adoption of any discount rate, however minimal, leads
   to short planning horizons. Rather than limit planning in this
   manner to a few decades at most, Doeleman argues for the
   adoption of absolute environmental standards.   Within this
   biologically determined framework, time preference can still
   operate, allowing for the relatively short planning horizons
   which humans are most capable of in day-to-day decision making.

194
Dorfman, Robert
   1976 <u>Forty Years of Cost-Benefit Analysis</u>.   Discussion Paper No. 498.
   Cambridge: Harvard Institute of Economic Research.

Dorfman offers a brief review of the forty-year history of B/C analysis as practiced in the United States. He discusses the original rationale, the limitations of the technique, and subsequent efforts to overcome these limitations. Present efforts have resulted in improvements in the technique, yet Dorfman argues that, since B/C analysis is based on doubtful welfare functions, inherent limitations remain.

195
Eckstein, Otto
    1968   Economic Analysis of Public Investment Decisions: Interest
           Rate Policy and Discount Analysis.   Report of the Subcommittee
           on Economy in Government of the Joint Economic Committee.
           Ninetieth Congress, Second Session.   Washington, D.C.:   U.S.
           Government Printing Office.

           The applications of discount rates in B/C analysis are
           neither adequate nor consistent.   Eckstein recommends the
           uniform application of discount rates equal to the opportunity
           cost of displaced private spending.

196
Fischer, David W.
    1970   "The problems of public investment criteria--A critique."   The
           Engineering Economist 15 (Winter): 123-127.

           Fischer appraises a study by Stephen A. Manglin entitled
           Public Investment Criteria: Benefit-Cost Analysis for Planned
           Economic Growth, Cambridge: MIT Press (1967).   He finds this
           suggested policy instrument lacking in internal consistency and
           applicability.

197
Fischhoff, Baruch
    1977   "Cost benefit analysis and the art of motorcycle maintenance."
           Poultry Sciences 8 (June):   177-202.

           Fischhoff offers a critical overview of decision tools
           (generally labeled cost-benefit analysis by the author).
           Discussed are: (1) their policy rationales; (2) their accepta-
           bility as guides to decision making; (3) problems encountered
           in their application; (4) ways in which they are misused; and
           (5) what steps must be taken to improve their contribution to
           society.   While the author finds flaws and weaknesses in cost-
           benefit approaches, he feels they will continue to play a
           critical role in the evaluation of new technology.   For BCA to
           play a useful role in decision making, analysts and decision
           makers must be aware of the weaknesses.   Fischhoff urges the
           institutionalization of more critical perspectives in the
           application of BCA.

198
Foster, John H.
    1976   "Flood management: Who benefits and who pays."   Water
           Resources Bulletin 12 (October): 1029-1039.

The Flood Control Act of 1936 established B/C analysis as
the basic decision-making tool for federal investment in flood
control.   It explicitly made the distribution of those costs
and benefits legally irrelevant. Thus, federal policy has led
to construction of flood control structures which tend to shift
costs of protection from owners of floodplains to the general
taxpayer. The author examines the alternatives to structure
building--disaster relief, flood proofing, watershed management, floodplain zoning, and flood insurance--and finds that
several of these internalize costs associated with floodplain
use.   Incorporating this distributive effect in the analysis
leads to more socially optimal flood management techniques.

199
Fox, Irving K., and Orrise Herfindahl
  1964 "Efficiency in the use of natural resources." American
  Economic Review 54 (May): 198-206.

This paper states that inefficient allocation of resources
in Corps of Engineers projects results from (1) zero or nominal
pricing of output which distorts demand, and (2) biased
practices in B/C analysis. These biases include inaccuracy of
estimates of economic parameters, and unreliable data and
questions of project size related to benefits. Another result
is that significant alternatives for satisfying water demand
are not considered. These problems can be solved only by
implementing independent audits of planning activities.

200
Freeman, A. Myrick III
  1969 "Income redistribution and social choice: A pragmatic
  approach." Public Choice 7 (Fall): 3-22.

Freeman develops a rationale and formula for determining a
social welfare function. This procedure is based on societal
preferences as revealed by tax, transfer, and expenditure
policy that alter the distribution of income. This welfare
function is then useful in B/C analysis of policies and
projects that have redistribution as one of their goals.

201
Freeman, A. Myrick III
  1967 "Six federal reclamation projects and the distribution of
  income."   Water Resources Research 3 (Second Quarter):
  319-332.

Following the observation of others that B/C ratios
produced by the Bureau of Reclamation are inflated, Freeman
uses available data to arrive at adjusted ratios for six recent
projects. While Bureau estimates of B/C ratios were all
greater than one, the author's estimates for five of the six
were less than one. Next, the hypothesis that redistribution
was the actual goal of these projects was tested by adjusting
the analysis for increases in equity resulting from the
projects. All projects resulted in an increase in equity;
however, the small size of the redistribution coupled with the

large cost of achieving it resulted in no significant change in
B/C ratios. Freeman concludes that other extra-economic goals
must have been fulfilled by these projects.

202
Freeman, A. Myrick III
   1966 "Adjusted benefit-cost ratios for six recent reclamation
   projects." Journal of Farm Economics 48 (November): 1002-1012.

      This study examines B/C estimates computed in six Bureau
      of Reclamation water projects. The author identifies major
      sources of bias in the process used and adjusts the ratios
      where data are readily available. The principal identified
      sources of error were 1) using artificially low discount rates,
      2) counting secondary benefits, 3) using overly optimistic
      price projections, 4) not counting the opportunity cost of farm
      labor, and 5) not counting the opportunity cost of water. Once
      adjustments were made for these factors, the B/C ratios of five
      of the six projects fell below one.

203
Freeman, A. Myrick III, and Robert H. Haveman
   1970 "Benefit-cost analysis and multiple objectives: Current issues
   in water resources planning." Water Resources Research 6
   (December): 1533-1539.

      The correct application of B/C analysis demands that 1)
      costs and benefits be quantifiable and that 2) quantification
      must be in the same units for all benefits and costs. One
      approach to analysis when one or both of these conditions is
      violated is a technique called multiple objective analysis as
      presented by David C. Major, "Benefit-cost ratios for projects
      in multiple objective investment programs," Water Resources
      Research 5 (1969): 1174-1178. The authors argue that Major's
      empirical and analytic foundations are weak and, further, that
      economists are not likely to solve these empirical and theo-
      retical difficulties in the near future.

204
Graaf, J. De V.
   1975 "Cost-benefit analysis: A critical view." South African
   Journal of Economics 43 (June): 233-244.

      The author examines the logical structure of B/C analysis
      and finds several internal inconsistencies. However, this
      methodology does take a wider view than conventional profit-
      and-loss accounting. The author urges keeping its flaws in
      mind when deciding whether to employ it as a decision-making
      tool.

205
Green, Mark, and Norman Waitzman
   1980 "Cost, benefit, and class." Working Papers 7 (May/June):
   39-51.

This article examines the class biases produced by the theory and application of B/C analysis. The authors' analysis suggests that these biases produce B/C analyses that emphasize benefits to the wealthy while downplaying the impact of costs borne by the poor and the working class.

206
Green, Harold P.
1975 "The risk-benefit calculus in safety determinations." George Washington Law Review 43 (March): 791-807.

Green responds to a speech made in 1973 before a National Academy of Sciences Forum by Philip Handler concerning the process of making safety determinations. Green contrasts his view as a lawyer to that of Handler, a scientist, and concludes that the formation of public policy should not be made by experts alone. Scientists must work within traditional policy-making institutions so that nonquantifiable values may be brought into the decision-making process. Handler's rebuttal follows.

207
Hammond, Richard J.
1966 "Convention and limitation in benefit-cost analysis." Natural Resources Journal 6 (April): 195-222.

This paper examines B/C analysis as defined by the federal government's guidelines for evaluating water resource projects. While B/C analysis is proposed as a method for objectifying the decision-making process, the author argues that, as practiced, it is a system of assumptions that are not based on objective logic. Rather, it is based on value judgments and should be recognized for what it is--a useful way of roughly estimating project outcomes and comparing the utility among competing projects. He urges that impersonal calculation not be used to replace responsible personal decision making.

208
Hanke, Steve H.
1981 "On the feasibility of benefit-cost analysis." Public Policy 29 (Spring): 147-157.

Correct application of B/C analysis requires that consistency between B/C measures and policies of rationing supply and acquisition must be maintained. The technical difficulties of maintaining this condition in the public sector are discussed, as are the political obstacles associated with adopting policies which remove these difficulties. Given these problems, the author questions the feasibility of correctly conducting B/C analysis.

209
Hanke, Steve H., and Richard Walker
1974 "Benefit-cost analysis reconsidered: An evaluation of the Mid-State Project." Water Resources Research 10 (October): 898-909.

The authors illustrate the misuse of B/C analysis by showing how adopting different assumptions leads to divergent
decisions regarding a proposed water project. They argue that
the selection of appropriate projects is fundamentally a political process.   Since B/C analysis is based on politically
derived assumptions, its use cannot replace political with
technical criteria in the public choice arena.   B/C analysis
should play a reduced role in public decision making and, where
employed, it should be kept simple so that its assumptions can
be easily identified.

210
Hanke, Steve H.
    1973   "The   political   economy   of   water   resources   development."
    In Transactions of the North American Wildlife and Natural
    Resources Conference 38, March 18-21, pp. 377-389. Washington,
    D.C.: Wildlife Management Institute.

        Hanke focuses on the consequences of the "evaluation-reim-
    bursement" dichotomy in federal water projects.   The author
    argues that, in general, B/C analysis results in exaggeration
    of demand for water projects since the cost of the water is
    often externalized to federal taxpayers.   This results in
    overdesigning and premature investment in facilities.   This
    then distorts the B/C ratio of future analyses since cost of
    services is not reflected. Evaluation and pricing, the author
    submits, should be integrated in analysis and, as a pricing
    rule, project beneficiaries should pay incremental costs.

211
Harberger, Arnold C.
    1979   "On the use of distributional weights in social cost-benefit
    analysis." Journal of Political Economy 86 (April Supplement):
    S.87-S.121.

        Harberger offers a methodology for determining distribu-
    tional weights for answering equity concerns in B/C analyses.
    Demand and supply elasticities are central to this method.   The
    impact on the results of BCA is illustrated for investment
    projects and in solving rudimentary tax problems.
        The use of distributional weights has drawbacks.   When
    even modest transfers of wealth result from the use of
    distributional weights, economic efficiency will decline, under
    some conditions rather dramatically.   Further inefficiencies
    may result from the costs of effecting the transfers. These
    trade-offs between efficiency and equity may be desirable, but
    Harberger argues that limiting economic analysis to traditional
    economic efficiency criteria would reflect a more modest and
    realistic appreciation of the role of economic analysis in
    decision making.

212
Harrison, David Jr.
    1975   Who Pays for Clean Air?:   The Cost and Benefit Distribution of
    Federal Automobile Emission Standards.   Cambridge, Mass.:
    Ballinger.

Harrison's analysis departs from the usual study of the
economic impact of a program. Instead, he concentrates on the
impacts for different economic groups. He finds that most of
the federal automobile standards are significantly regressive.
An alternative scheme for controlling pollution is presented
which is more cost-effective than those proposed by the federal
standards and which avoids undesirable distributional effects.

213
Haveman, Robert H.
   1972 The Economic Performance of Public Investment: An Ex-Post
         Evaluation of Water Resources Investment. Baltimore: Johns
         Hopkins University Press.

         While cost-benefit analysis is increasingly used in deci-
         sion making on public expenditures, little postconstruction
         investigation has been conducted to determine the accuracy of
         projected efficiency. The book evaluates the agreement between
         projected results and consequences of several federal water
         resource investments. The author finds a systematic over-
         estimate of expected benefits and large variance between esti-
         mated and realized costs.

214
Haveman, Robert H.
   1969 "The opportunity cost of displaced private spending and the
         social discount rate." Water Resources Research 5 (October):
         947-956.

         Haveman describes major schools of thought on how the
         social interest rate (or discount rate) should be set when
         doing economic analysis. Economists have been unable to arrive
         at consensus on which method is appropriate or on how to
         delineate the divergent value judgment of each position. While
         the author argues for the adoption of his method for setting
         the discount rate, his own argument leads to the conclusion
         that there is no best solution. The discount rate must be
         defined in accordance with personal or societal values.

215
Herschaft, Alex, A. Myrick Freeman III, Thomas D. Crocker, and Joe B.
Stevens, eds.
   1978 Critical Review of Estimating Benefits of Air and Water Pollu-
         tion Control. Report prepared for the U.S. Environmental Pro-
         tection Agency, Office of Health and Ecological Effects and the
         Office of Research and Development. Available from National
         Technical Information Service as PB-285 555/9SL.

         This report, a critical review of the state of the art and
         future prospects for estimating the benefits of air and water
         pollution control, addresses the benefit side of B/C analysis.
         Specific topics covered are the nature and role of benefits,
         damage functions, valuation of effects, aggregation of results,
         and the handling of uncertainties. The conceptual basis for
         benefit estimation is presented and compared to actual empiri-
         cal studies. The authors conclude that benefit estimates do

not usually reflect the state of the art but that pollution control benefit estimation is potentially useful to decision makers.

216
Hettich, Walter
    1976 "Distribution in benefit-cost analysis: A review of theoreti-
          cal issues." Public Finance Quarterly 4 (April): 123-150.

          After discussing the basic theoretical framework of
    benefit-cost analysis, welfare economics, and the measurement
    of income equivalents, Hettich presents a review of several
    attempts to construct distributional weights for the integra-
    tion of equity and efficiency in evaluation. Also reviewed are
    several critiques of attempts to include equity considerations
    in benefit-cost analysis. Hettich argues that these objections
    are not valid and concludes that there is little theoretical
    reason for disregarding equity goals in analysis. He admits,
    however, that there is as yet no consensus as to how distribu-
    tional goals should be incorporated into B/C analysis.

217
Hettich, Walter
    1971 Why Distribution Is Important: An Examination of Equity and
          Efficiency Criteria in Benefit-Cost Analysis. Ottawa:
          Information Canada.

          This monograph discusses the rationale for integrating
    equity and efficiency dimensions in B/C analysis. It explores
    methods for doing this in Canadian policy making.

218
Hirsheifer, Jack, James C. DeHaven, and Jerome W. Milliman
    1960 Water Supply: Economics, Technology and Policy. Chicago:
          University of Chicago Press.

          This book discusses the application of economic and tech-
    nological knowledge to the solution of water resource prob-
    lems. In large part, the authors are cautious about the bene-
    fits of water development projects. Their analysis shows that
    sound economic principles are often ignored in the analysis of
    proposed projects. Consequently, uneconomic structures are
    built and alternative management techniques are ignored. The
    cause of this is, according to the authors, that personal
    rewards associated with administering the construction of large
    projects are greater than those associated with advocating more
    efficient utilization of resources at hand. The book is
    therefore relevant in interpreting psychological states and
    personal interests underlying B/C analysis.

219
Joksch, H. C.
    1975 "A critical appraisal of the applicability of benefit-cost
          analysis to highway traffic safety." Accident Analysis and
          Prevention 7: 133-153.

B/C analysis of highway traffic safety measures is widely practiced. The author develops the idea that the acceptance of this methodology is not based on thorough examination of the conceptual and practical problems arising when B/C analysis is applied to the study of traffic safety. This study critically reviews the assumption of B/C analysis and illustrates the practical problems of estimating benefits and costs. A limited scope for the appropriate application of benefit-cost analysis is defined.

220
Jones-Lee, M. W.
    1976   The Value of Life: An Economic Analysis. Chicago: University
            of Chicago Press.

        This book takes as its purpose the discovery of a proce-
    dure for placing a value upon the saving of human life or
    avoidance of human injury. It is at once philosophical and
    mathematical but quite accessible to readers seeking a review
    of "value of life" literature. The seven chapters are
    summarized in the preface:
        "Chapter 1 presents a brief outline of the conventional
    procedures of public sector allocative decision-making, special
    consideration being given to the fundamental rationale for
    cost-benefit analysis. Chapter 2 surveys the existing litera-
    ture in the 'value of life' and 'safety improvement' field and
    evaluates each contribution with particular reference to its
    relevance for cost-benefit analysis. Chapter 3 contains a
    brief exposition of those aspects of the theory of choice under
    uncertainty that are particularly important for later
    chapters. Chapter 4 examines the individual's decision con-
    cerning expenditure on life insurance, emphasis being placed
    upon the necessary and sufficient conditions for the purchase
    of such insurance to be desirable from the individual's point
    of view. These conditions then play an important part in the
    development, in chapter 5, of a number of qualitative results
    concerning the value of changes in safety and longevity.
    Chapter 6 outlines an experimental procedure for obtaining
    quantitative estimates of a limited number of such experi-
    ments. Finally, chapter 7 surveys the arguments of the
    preceding chapters, explores strengths and weaknesses and
    offers suggestions for future research."

221
Kazis, Richard, and Richard L. Grossman
    1982   Fear at Work. New York: The Pilgrim Press.

        This work could easily fall within either set of annota-
    tions in this volume. Born of the work and research of
    Environmentalists for Full Employment (EFFE), it is a compre-
    hensive chronicle of ways in which the business community, with
    government acquiescence, uses "job blackmail" (that is, dissem-
    ination of the argument that the nation can have either jobs or
    environmental quality, but not both) to turn working people
    against environmental reform and regulation. The authors offer
    extensive documentation defending a counterhypothesis--that

environmental protection for citizens and workers alike creates
new jobs and saves the business community substantial sums in
unpaid compensation and sick days. Many jobs "lost" due to
stricter regulation were, in their view, slated for extinction
by industries eager to leave an area or to replace workers with
new labor-saving technologies.

Unlike many authors who share this view, Kazis and
Grossman construct their case from the larger political economy
of the United States. The deeper roots of recent job reductions and layoffs are traced to untenable growth ideals of the
postwar decades, the inflation legacy of Vietnam, and national
economic policies in the wake of the 1974 recession and energy
crisis which thrust the costs of recovery onto American
workers. Particular attention is paid to deregulatory drives
spearheaded by the business community in the 1970s which echoed
tellingly in the Reagan election. Chapters 8 ("Innovation and
Environmental Regulation") and 9 ("Is Environmental Protection
Worth the Cost?") come to terms frontally with the BCA challenges underlying deregulation on the eve of and during the
Reagan administration. The book ends with a thoughtful appeal
for further labor-environmental organizing efforts.

222
Kelman, Steven
   1981   "Cost-benefit analysis: Ethical critique."   Regulation 5
          (January/February):   33-40.

          In the context of formal theory, Kelman examines the basic
propositions of B/C analysis. He concludes that (1) in areas
of environmental, safety, and health regulation, there may be
instances when a decision might be right even though its costs
outweigh its benefits; (2) there are good reasons to oppose the
dollar valuation of nonmarket benefits and costs necessary for
the application of B/C analysis; and (3) given these
limitations, devoting significant resources to the collection
of data necessary to perform B/C analysis or to promote
widespread use of the method cannot be justified.

223
Kendall, Henry, and Sidney Moglewer
   1975   "Preliminary review of the AEC reactor safety study."   In
          Benefit-Cost and Policy Analysis 74, ed. Richard Zeckhauser,
          Arnold A. Harberger, Robert H. Haveman, Laurence E. Lynn, Jr.,
          William Niskanen, and Alan Williams, pp. 66-80.   Chicago:
          Aldine Publishing Company.

          This paper is excerpted from a larger study of the same
title prepared under the direction of the Sierra Club and the
Union of Concerned Scientists. It critiques the Rasmussen
"Reactor Safety Study." The combined underestimation of acci-
dent probabilities and of the consequences of those accidents
results in assessment of risk which is underestimated by
several orders of magnitude. If revised estimates of risk are
entered into B/C analysis, the expected costs exceed expected
benefits.

224
Kneese, Allan V.
   1973 "The Faustian bargain." Resources 44 (September): 1-5.

          Kneese believes that B/C analysis cannot answer the most
      important policy questions regarding the desirability of a
      large-scale, nuclear-based economy.   There are ethical
      questions associated with risks which cannot be solved by B/C
      analysis; indeed B/C analysis may obscure those questions.

225
Krutilla, John V., and Anthony C. Fisher
   1975 The Economics of Natural Environments.   Baltimore: Johns
      Hopkins University Press.

          This book presents methods enabling the integration of
      nonmarket or amenity values of natural environments into the
      theory and practice of economics, values traditionally ignored
      in B/C analysis.

226
Krutilla, John
   1961 "Welfare aspect of benefit-cost analysis." Journal of Politi-
      cal Economy 69 (April): 226-235.

          Krutilla examines B/C analysis in the context of the
      theory of welfare economics. He finds that the method is par-
      tially contradictory with this body of theory. However, his
      conclusion is that using B/C analysis as a decision-making tool
      is preferable to decision making with no analysis at all.

227
Kula, E.
   1980 "Future generations and discounting rules in public sector
      investment appraisal." Environment and Planning 13 (July):
      899-910.

          When a discount rate is applied to public investment that
      has long-term benefits, these benefits to future generations
      will be discounted to zero. This discriminates against pos-
      terity as an artifact of the methodology used. The author
      presents an alternative methodology, the "sum of discounted
      consumption flows," that allows the analysis to treat present
      and future generations equally.

228
Lind, Robert C.
   1982 Discounting for Time and Risk in Energy Investment Planning.
      Baltimore: Johns Hopkins University Press.

          A renewed debate on the appropriate discount rate emerges
      in the context of proposed large-scale energy development
      projects. Past work on the selection of a rate of discount
      has centered on the economic evaluation of large water projects
      in the late 1950s and 1960s. Important differences exist be-
      tween the context of the former theoretical development and the
      current situation, however. First, economists have in the past
      been influenced by the prospect of ever-increasing per capita

wealth of future generations; evaluators of present-day projects must be less optimistic on this point. Second, previous
development of appropriate rates of discount was based on a
perceived trade-off between more consumption now and relatively
less consumption in the future. Now, the perceived trade-off
is often seen in terms of relatively more consumption today and
catastrophic costs in the future. These differences have
forced a reevaluation of discount rates.

Lind has collected in this volume a series of commissioned
papers addressing the unresolved issues related to the choice
of discount rate to be used in B/C analysis of public investment and public policies related to large-scale energy projects. Among the issues addressed are the question of risk in
discounting, whether discounting at any rate implies intergenerational equity, and whether BCA is appropriate where serious
intergenerational consequences exist. Practical problems
associated with the estimation of an appropriate rate are
addressed; finally, an evaluation of discounting practices
within the electric power industry is presented.

229
Liroff, Richard A.
    1982 "Cost-benefit analysis in federal environmental programs." In
    Cost-Benefit Analysis and Environmental Regulations: Politics,
    Ethics and Methods, ed. Daniel Swartzman, Richard A. Liroff,
    and Kevin G. Croke. Washington, D.C.: The Conservation
    Foundation.

        Liroff gives the political history of executive order and
    statutory requirements for cost-effectiveness in federal regu-
    lations. The effects of these precedents are then extrapolated
    to BCA in environmental regulations. Liroff notes that the EPA
    has a history of reporting the benefits and costs of its
    actions which are not, however, translated into cost-
    benefit analyses. Doing BCA may, in Liroff's view, improve
    economic efficiency and strengthen support for environmental
    regulations as long as EPA can dissociate its analyses from
    analyses done by the industries which it regulates.

230
Maas, Arthur
    1966 "Benefit-cost analysis: Its relevance to public investment
    decisions." Quarterly Journal of Economics 80 (May): 208-226.

        B/C analysis as applied to public investment in the United
    States is capable of ranking projects and programs only in
    terms of their economic efficiency. Goals other than effi-
    ciency may be important, even primary: redistribution and
    promotion of natural self-sufficiency are examples. Political
    institutions should be used to measure the trade-off between
    efficiency and important nonefficiency objectives of a public
    project to explicitly take into account other societal goals.

231
MacMillan, James A.
    1976 "A critique of Benefit-Cost Analysis Guide, Planning Branch,
    Treasury Board Secretariate. Ottawa: Information Canada."
    Canadian Journal of Agricultural Economics 24 (November):
    50-54.

MacMillan critiques the guide in terms of inconsistencies and errors as well as for the misapplication of B/C analysis advocated. He concludes that B/C analysis should be used only to rank similar projects and that economic analysis for the Canadian federal government should concentrate more on income distribution effects of public policy.

232
Maclean, Douglas
  1980 "Benefit-cost analysis, future generations and energy policy: A survey of moral issues." Science, Technology and Human Values 5 (Spring): 3-10.

       Present decision making frequently affects future genera-tions insofar as it commits society to a particular path of development. Traditional B/C analysis fails to adequately take into account these intergenerational costs and benefits. Maclean treats the problems of discount rate, of distribution effects of costs and benefits across generations, and of reducing difficult decision making to a "unified but artificial system...." He then presents a philosophical discussion of intergenerational obligations.

233
Mac Rae, Duncan, Jr.
  1976 The Social Function of Social Science. New Haven: Yale University Press.

       Particularly in chapter 5 of this book, three basic types of economic ethics are considered: (1) "old" welfare econo-mics, (2) its successor, the "new" welfare economics, which is based on Pareto optimality, and (3) B/C analysis with founda-tions in "new" welfare economics. The three types are compared and a general criticism based on the distinction between preference and welfare is developed.

234
Mark, R. K., and D. E. Stuart-Alexander
  1977 "Disasters as a necessary part of benefit-cost analyses." Science 197 (16 September): 1160-1162.

       B/C analyses for water projects commonly do not include expected costs related to low-probability events like dam failures. Analysis of the history of such events suggests that dam failures are not uncommon. Since expected costs from such events can be significant, estimates of these costs should be included for each project. Their omission results in an upward bias in the cost-effectiveness estimates of projects.

235
Marsh, Georganne E.
  (1971-present) Benefit-Cost and Policy Analysis. Chicago: Aldine Publishing Co. (annual series).

       This annual series, initiated in 1971 as Benefit-Cost Analysis, changed to its current title in 1972. The series attempts to select the most pertinent articles on the topic from a range of scholarly and applied sources. Its relevance to the present bibliography is twofold: (1) articles appear which challenge as well as accept the overall logic and method of B/C analysis, and (2) articles presenting important extensions and corrections of the technique are routinely included.

236
Masters, Stanley, Irwin Garfinkel, and John Bishop
    1979 "Benefit-cost analysis in program evaluation." Journal of Social Service Research 2 (Fall): 79-93. (Also reprinted by Institute for Research on Poverty, University of Wisconsin-Madison, Reprint Series No. 410)

BCA, the authors argue, is nothing more than the careful comparing of the costs and benefits of alternative programs of activities. That both costs and benefits are not readily measured and may be evaluated differently by different interested parties necessitates extreme care in the conduct of BCA if it is to prove useful. Such analysis should be done with imaginative reckoning of costs and benefits by people knowledgeable about the proposed project. Nonquantifiable values should be included in the analysis by giving explicit attention to policy implications of unmeasured value judgments. To aid users of a BCA, the analyst's own value judgments and their policy implications should be stated explicitly.

237
Mathur, Vijay K.
    1971 "The integration of equity and efficiency criteria in public project selection: A comment." Economic Journal 81 (December): 929-931.

Mathur presents a brief critique of McGuire and Garn's (1969) article. He addresses several issues: (1) the claim that area selection is tantamount to project selection, (2) the assumption that rich areas are more efficient than poor areas in the production of public goods, and (3) the value judgment that an explicit, as opposed to implicit, weighting method results in better project choice criteria. In addition, Mathur discusses income and employment variables omitted by McGuire and Garn. Mathur's conclusion is that there is no analytical difference between explicit and implicit weighting systems. Implicit weighting is, however, much simpler to apply in practical applications.

238
McGuire, Martin C., and Harvey A. Garn
    1969 "The integration of equity and efficiency criteria in public project evaluation." Economic Journal 79 (December): 882-893.

Following recent efforts of others calling attention to the previously ignored equity consideration in project evaluation, McGuire and Garn present formulas for the integration of equity and efficiency criteria in regional development projects. They base their work on an index of need reflecting marginal utility of development for communities. This index is then used as a weight in cost-benefit analysis to reflect distributional goals.

239
McKean, Roland N.
  1958 Efficiency in Government Through Systems Analysis: With Emphasis on Systems Analysis. New York: John Wiley & Sons, Inc.

         McKean presents general methodological problems in the application of B/C analysis to water resources projects and other areas of development in which B/C analysis might improve the efficiency of public investment.

240
Merewitz, Leonard
  1973 "Cost overruns in public works." In Benefit-Cost and Policy Analysis 72, ed. William A. Niskanen, Arnold C. Harberger, Robert H. Haveman, Ralph Turvey, and Richard Zeckhauser, pp. 277-295. Chicago: Aldine Publishing Company.

         Sizable cost overruns on public projects are commonplace. This does not appear to have hampered expansion of public sector investment, however. Recent studies suggest that these overruns can be estimated and predicted. An unfortunate side effect of anticipating overruns is laxity in cost control.

241
Miller, James C., III, and Bruce Yandze, eds.
  1979 Benefit-Cost Analyses of Social Regulation. Washington, D.C.: American Enterprise Institute for Public Policy Research.

         This volume summarizes BCA case studies by economists on the staff of the Council on Wage and Price Stability. Topics analyzed are divided by the editors into three broad categories: (1) consumer and worker health; (2) product safety; and (3) energy, environment, and international trade. The editors offer broad conclusions which may be drawn from these studies: (1) legislation often constrains the extent to which BCA is taken into consideration in promulgating regulations; (2) agencies tend to substantially reflect the views of the constituents that supported their formation; and (3) BCA monitors public sector efficiency in much the same way that competition contributes to private sector efficiency.

242
Mishan, Edward J.
  1976 Cost-Benefit Analysis. 2nd ed. New York: Praeger Publications.

         This book is among the most comprehensive introductions to the subject of B/C analysis. It will be of use to those seeking a clear, nonmathematical presentation of its basic precepts and procedures. The discussion is accessible to readers with only a smattering of formal economic training.

243
Mitchell, Bruce, and Joan Mitchell
    1972 Benefit-Cost Analysis: A Select Bibliography. Monticello,
          Ill.: Council of Planning Librarians.

          This extensive bibliography of BCA literature up to 1972
          is divided into two sections. Part I cites background
          information on BCA and cost-effectiveness analysis; Part II
          gathers articles illustrating the application of BCA methods to
          a variety of areas. Several articles found in this biblio-
          graphy are listed separately in the present work.

244
Moody, R. E.
    1974 "Inadequacy of the cost-benefit ratio as a measure of the
          public interest." American Journal of Agricultural Economics
          56 (February): 188-191.

          Moody argues that while the B/C ratio is based on empiri-
          cally established valid indicators of private interests, it is
          not established as an index of public interest. His contention
          is that the structures of private and public interest are so
          different that B/C analysis cannot be transferred wholesale
          from one to the other.

245
Müller, Frank G.
    1974 "Benefit-cost analysis: A questionable part of environmental
          decisioning." Journal of Environmental Systems 4 (Winter):
          299-307.

          Müller suggests that environmental health requires a level
          of economic activity which maintains ecological stability. The
          optimal level of resource exploitation suggested by BCA, on the
          other hand, is determined by the Pareto-optimal condition,
          without reference to physical and biological facts necessary
          for an assessment of the ecological health of the environment.
          Utilizing BCA in the evaluation of environmental policy there-
          fore can, and often does, result in either misleading or super-
          fluous information.

246
Nash, Christopher, David Pearce, and John Stanley
    1975 "An evaluation of cost-benefit criteria." Scottish Journal of
          Political Economy 22 (June): 121-134.

          The authors argue that there are many ways of performing
          B/C analysis and that each is logically consistent within a
          particular set of moral values. The choice of method is an
          ethical one. To be most useful as an input to public debate
          and decisions, presentation of B/C results should be
          disaggregated to show the incidence of costs and benefits on
          affected groups and should clearly show the method of valuation
          used.

247
Nath, S. K.
    1969 A Reappraisal of Welfare Economics.   Fairfield, N.J.: August
          M. Kelley Publishers.

          Nath examines and questions the logic of some of the
          commonly accepted propositions of welfare theory (the theoreti-
          cal basis of B/C analysis). The author concludes that because
          of its limitations, a priori application of the theory is
          inappropriate. However, if the value judgments employed in a
          given analysis are made explicit, welfare economics can be
          fruitfully employed.

248
Otway, Harry J., ed.
    1972 Risk vs. Benefit: Solution or Dream. Compendium of papers
          from a symposium sponsored by the Western Interstate Nuclear
          Board. Los Alamos, NM: Los Alamos Scientific Laboratory.

          This collection of papers contains discussion of risk-
          benefit analysis as applied to nuclear energy. Of particular
          interest is H. P. Metzger's popularized account of problems
          related to such analysis, "Benefit-Risk Analysis Is a Dream,"
          and H. J. Otway's "The Quantification of Social Values." Paul
          Slovic, in "Limitations of the Mind of Man: Implications for
          Decision Making in the Nuclear Age," discusses the problems
          trained scientists have in making rational decisions concerning
          processes that are of a probabalistic nature.

249
Peacock, Alan
    1973 "Cost-benefit analysis and the political control of public
          investments." In Cost-Benefit and Cost Effectiveness: Studies
          and Analyses, ed. J. N. Wolf, pp. 17-23. London: George Allen
          and Unwin.

          Peacock discusses the value judgments embodied in B/C
          analysis of government investments. He believes that for such
          analysis to be valid, these value judgments should be made
          explicit to--and be approved by--the relevant decision-making
          body.

250
Pearce, David
    1976 "The limits of cost-benefit analysis as a guide to environ-
          mental policy." Kyklos 29: 97-112.

          This paper defines certain limitations of B/C analysis as
          a criterion guiding pollution control.   B/C analysis is shown
          to be of limited value when (1) pollution has sustained
          biological effects, (2) the pollutant has biological effects
          and the assimilative capacity of the environment is low and,
          (3) when increasing levels of conventional pollutants lead to
          dynamic instability of the ecosystem. Pearce concludes that
          the "true" optimal pollutant level in these cases is less than
          that dictated by B/C analysis. The article is critiqued in

Kyklos 30 (1970) by David Elliot and George Yarrow, pp. 300-309, and by V. Kerry Smith, pp. 310-313; Pearce responds to these remarks on page 314-318.

251
Pingle, Gautam
   1978   "The early development of cost-benefit analysis." Journal of Agricultural Economics 29: 63-71.

   Pingle traces the inception of B/C analysis to 1760s' Scotland and finds that early methodological sophistication reached a height in the mid-nineteenth century among British officials in India. The main impetus appears to have come from the need of civil servants to convince a skeptical home government of the need to develop agricultural infrastructure. Thus, the analysis provided justification for projects rather than tools for investment planning. Pingle argues that the situation is little changed today and urges the comprehensive identification and inclusion of investment alternatives in analysis as a way of avoiding this problem.

252
Prest, A. R., and R. Turvey
   1965   "Cost-benefit analysis: A survey."   Economic Journal 75 (December): 683-735.

   The article reviews the theory and practice of B/C analysis up to 1965:   its development and scope, its general principles, particular applications of the method, and the state of the art, including important limitations. An extensive bibliography concludes the article.

253
Ragade, R. K., Keith W. Hipel, and T. E. Unny
   1976   "Metarationality in benefit-cost analyses."   Water Resources Research 12 (October): 1069-1076.

   Traditional benefit-cost analysis is often criticized because no satisfactory treatment of intangible values is available, and because cost minimization and benefit maximization are objectives that may be viewed differently by various interest groups. To overcome these objections, the authors introduce a type of B/C analysis which is capable of handling quantitative and qualitative values and which allows each interest group to assign values to benefits and costs according to any criterion it chooses. A theory of linguistic descriptions is proposed in which metarationality is introduced and which uses pseudo-Boolean functions to define concepts and record preferences of parties involved in an analysis.

254
Renshaw, Edward F.
   1957   Toward Responsible Government. Chicago: Idyia Press.

   Renshaw is critical of the economic parameters employed in B/C analysis of large, federally funded water projects. Using what he considers more realistic procedures, he shows that most projects are uneconomic. The pressures for building these projects, he explains, derive for the most part from the fact that beneficiaries do not bear the costs of project outputs.

255

Ruttenberg, Ruth

1981 "Why social regulatory policy requires new definitions and techniques for assessing costs and benefits: The case of occupational safety and health." Labor Studies Journal 6 (Spring): 114-123.

While analyses of social policy relevant to labor should include a rigorous accounting of advantages and disadvantages, their reduction to the common metric of dollars required by B/C analysis is inappropriate. In addition, distributional effects, which may be of paramount importance, are not adequately handled in a B/C framework. Ruttenberg concludes that new multidisciplinary methodologies, in which economics plays but one part, must be developed for the analysis of social programs.

256

Ruttenburg, Ruth, and Randall Hudgins

1981 Occupational Health and Safety in the Chemical Industry. New York: Council on Economic Priorities.

This study presents an analysis of the impact of the Occupational Safety and Health Administration on eight major U.S. chemical companies. The study finds, contrary to criticisms of OSHA as being an excessively expensive regulatory burden, that the regulations have substantially reduced injury and illness rates at the modest cost of about $140 per worker per year. The study reports that the rate of violations per inspection in the chemical industry is nearly three times the national average and that the severity of violations exceeds all other industrial sectors except mining.

257

Sarrazin, Thio, Frithjof Spreer, and Manfred Reitzel

1974 "Theorie und Realität in der Cost-Benefit Analyse." ("Theory and reality in cost-benefit analysis," with English summary). Zeitschrift für die Gesamte Staatswissenschaft 130 (January): 51-74.

The authors examine the assumptions and techniques of B/C analysis, and find that the degree to which these accurately represent reality is limited in several important ways. They conclude that the usefulness of benefit-cost analysis as a decision-making tool is similarly limited.

258

Schelling, T. C.

1968 "The life you save may be your own." In Problems in Public Expenditure Analysis, ed. Samuel B. Chase, pp. 127-162. Washington, D.C.: Brookings Institution.

Schelling begins by discussing the social value of human life, and reviews several approaches to valuation of it. Given the unique aspects of human life, valuation is difficult, but Schelling offers a number of methods for overcoming some of the difficulties.

259
Schnaiberg, Allan, and Errol Meidinger
   1978 "Social reality versus analytic mythology: Social impact assessment of natural resource utilization." Paper presented at the annual meeting of the American Sociological Association, San Francisco, California.

In principle, the assessment of natural resource utilization and decision making related to that assessment are based on scientific management. The technique most frequently associated with such management is B/C analysis. However, political rather than scientific concerns predominate in public administration behavior. The analysis presented in this paper shows how B/C analysis in application reflects these political concerns rather than scientific-rational principles. The result is that economic rather than environmental interests have been favored in the assessment process. The authors suggest ways of redressing this bias.

260
Schnaiberg, Allan
   1980 "Social welfare intelligence." In Schnaiberg, The Environment: From Surplus to Scarcity, pp. 316-361. New York: Oxford University Press.

An earlier version of this chapter was presented in Schnaiberg and Meidinger 1978, "Social reality versus analytic mythology: Social impact assessment of natural resource utilization."

261
Self, Peter
   1970 "'Nonsense on stilts': Cost-benefit analysis and the Roskill Commission." Political Quarterly 41 (July/September): 249-260.

Self argues that B/C analysis is of limited value in decision making. Its limitations stem mainly from the valuation in monetary terms of goods that are essentially nonmonetary, the impossibility of including all important costs and benefits, and the difficulty of accounting for distributional effects. The result is to give spurious authority to what is in fact a point of view.
For a critique of this article and a rejoinder, see "Correspondence" from R. C. Fordham and the reply by Self in Political Quarterly 42 (January/March, 1971):94-95.

262
Self, Peter
   1975 Econocrats and the Policy Process: The Politics and Philosophy of Cost-Benefit Analysis. Boulder, Colorado: Westview Press, Inc.

Self (a noneconomist) presents a critical view of B/C analysis in a nonmathematical style. Part I scrutinizes the tools and methods of B/C analysis. Part II discusses the interrelationship between economic analysis, politics, social values, and methods of decision making.

263
Seneca, Joseph J.
1970 "The welfare effects of zero pricing of public goods." Public Choice 8 (Spring): 101-110.

Seneca argues that if public goals are not rationed by price, or if pricing is nominal, traditional B/C analysis will overstate welfare. Policy may result that does not utilize all available capacity with respect to welfare maximization.

264
Shabecoff, Philip
1981 "Reagan order on cost-benefit analysis stirs economic and political debate." New York Times, November 7: 28.

President Reagan's cost-benefit order of February 17, 1981, provided that "regulatory action shall not be undertaken unless the potential benefits to society from the regulation outweigh the potential costs to society." Shabecoff discusses several commentators' views of the advantages and disadvantages of the B/C analyses mandated by the order.

265
Sheff, Rosalyn L.
1979 Environmental Benefits Assessment in Economic Impact Studies: A Review. Chicago: Illinois Institute of Natural Resources. Available from National Technical Information Service as PB80-119985.

This is a retrospective review of environmental benefits assessments as practiced in the first five years of Illinois experience with economic impact studies. The author believes that such studies are valuable but do not go far enough. The interactions between pollution and the environment may result in costs beyond simple loss of environmental quality; the benefits of preventing these secondary damages should also be included in economic impact statements.

266
Shoup, Donald C.
1974 "Cost effectiveness of urban traffic law enforcement." In Benefit-Cost and Policy Analysis 73, ed. Robert H. Haveman, Arnold C. Harberger, Laurence E. Lynn, Jr., William A. Niskanen, Ralph Turvey, Richard Zeckhauser, and Daniel Wisecarver, pp. 188-213. Chicago: Aldine Publishing Company.

In the absence of information regarding the way various agency resources are converted to objectives, analysis of benefits and costs will contain major errors and lead to inefficient resource allocations. The chapter represents an attempt to develop an actual production function for social outputs for a specific agency with the aim of correcting this problem.

267
Shrader-Frechette, Kristin S.
1980 "Technology assessment as applied philosophy of science."
Science Technology and Human Values 6 (Fall): 33-50.

The central notions of B/C analysis are identified and the basic assumptions underlying each concept delineated. The author then identifies bias inherent in each assumption, along with probable consequences of those biases, and concludes that the application of B/C analysis necessarily involves certain value judgments.

268
Smith, R. Jeffrey
1981 "OSHA shifts direction on health standards." Science 212 (June 26): 1482-83.

This article catalogs agency changes made by the then newly appointed head of OSHA, Thorne Auchter. These changes effectively reduce health and safety standards in a number of sectors including the construction, chemical, and textile industries. Auchter suggests that his predecessor had pushed rules into areas where they were unworkable.
New agency policy will be to rely on BCA and not promulgate rules unless there is clear evidence that the benefits from additional regulations outweigh their cost. Another primary concern will be the workability of rules; regulations will not be instituted if OSHA feels they cannot be adequately enforced. In its efforts to redirect policy, OSHA is making increasing use of information provided by industry groups and relying less on the findings of its own technical staff. These new policy directions have drawn criticism from labor groups and OSHA staff members alike.

269
Spangler, M.B.
1980 United States Experience in Environmental Cost-Benefit Analysis for Nuclear Power Plants with Implications for Developing Countries. Washington, D.C.: Office of Nuclear Reactor Regulation, U.S. Nuclear Regulatory Commission (August). Available from National Technical Information Service as NUREG-0701.

Since the U.S. Circuit Court established in 1971 a requirement for an expanded Environmental Impact Statement (EIS) involving the balancing of environmental costs against benefits by the U.S. Atomic Energy Commission (AEC) and its successor, the Nuclear Regulatory Commission (NRC), the AEC/NRC has prepared over 140 EIS's. In regard to nuclear power plant licensing actions, environmental cost-benefit analyses relate to four types of alternatives: (1) the need for additional baseload electrical generating capacity versus the "no action" alternative; (2) alternative sources of energy for generating electricity; (3) alternative sites for a proposed nuclear power plant and its added transmission corridors; and (4) alternative technologies for the mitigation of adverse environmental impacts.

Spangler treats environmental impact assessment in two parts. Part I deals with Methods, Problems, and Issues in Environmental Cost-Benefit Analysis of Nuclear Power Plant Alternatives. Part II presents quantitative estimates for Material, Human, and Financial Resource Requirements for Nuclear Power Plant Construction and Operation in Comparison with Coal-Fueled Plants. The procedural and substantive problems and issues encountered by the United States in environmental cost-benefit analysis for nuclear power plants exhibit a number of important, if rough, parallels for other countries. Procedural issues include: (1) the identification of alternatives and related impact issues to be treated; (2) the problem of an environmental cost-benefit integration or balancing of incommensurate impacts stated in a mixture of quantitative and qualitative forms; and (3) methodologies for the treatment of uncertainty in measuring and forecasting certain kinds of environmental costs and benefits. Although substantive environmental issues will vary appreciably among nations, many of the substantive impact issues such as impacts on biota, community-related effects, and aesthetic impacts will also be of interest to labor groups outside the United States.

270
Strauch, Ralph E.
    1975 "'Squishy' problems and quantitative methods." Policy Science
          6 (June): 175-184

        'Squishy' problems--those which are not represented by a well-defined, unambiguous structure--present problems in the application of logical inference that underlies quantitative analytic methods. Strauch reasons that quantitative methodologies are not without value in policy analysis but that decision making should rely more on wise people than on sophisticated methods. Indirectly relevant to B/C analysis.

271
Sugden, Robert, and Alan H. Williams
    1978 The Principles of Practical Cost-Benefit Analysis. New York:
          Oxford University Press.

        The core exposition attempts to understand and thereafter to apply BCA rather than to criticize it. The authors attempt to make the economic assumptions and manipulations inherent in BCA elementary. To this end they offer simulation exercises at the end of each chapter.

        The exposition is most accessible to those with formal training in mathematics; technical appendixes appear along with guides after each chapter for further reading geared to economists and to BCA practitioners, including noneconomists. The final chapter and epilogue reflect on BCA as an analytic tool for community decision makers mindful of their broad public charge rather than just narrower programmatic ends.

272
Swartzman, Daniel
  1982 "Cost-benefit analysis in environmental regulation: Sources of
       the controversy." In Cost-Benefit Analysis and Environmental
       Regulations: Politics,   Ethics   and   Methods,   ed. Daniel
       Swartzman, Richard A. Liroff, and Kevin G. Croke. Washington,
       D.C.: The Conservation Foundation.

           Swartzman encapsulates arguments for and against BCA in
       the form of a dialogue. This refreshing approach serves as the
       point of departure for an examination of the BCA controversy.
       Swartzman argues that the sources of disagreement have not been
       adequately described.   He then proposes a taxonomy using
       differences in methodology, political views, and ethical
       beliefs as sources of controversy. Each of these sources is
       then dissected to show that BCA is consistent with distinct
       sets of assumptions, values, and beliefs. He stresses in his
       conclusion that he is not trying to resolve the BCA contro-
       versy; rather, he is trying to aid communication between the
       several sides of the debate. He suggests that fundamental
       differences among BCA defenders and detractors have led to
       difficulties in advancing the debate.

273
Tabb, William K.
  1980 "Government regulations: Two sides of the story." Challenge
       23 (November/December): 40-48.

           This article critiques a report by Murray L. Weidenbaum
       and Robert De Fina entitled The Cost of Federal Regulation of
       Economic Activity (Washington: American Enterprise Institute,
       1978).   Tabb argues that (1) contrary to the cited report's
       findings, recent social regulations like OSHA and NEPA
       contribute relatively small amounts to the total cost of feder-
       al regulation, (2) the costs of federal regulation and compli-
       ance costs reported in the original study are exaggerated, and
       (3) while the Weidenbaum and De Fina study did not report bene-
       fits of social regulatory programs, several studies suggest
       they are substantial.

274
Tribe, Laurence H.
  1973 "Technology assessment and the fourth discontinuity: The
       limits of instrumental rationality." Southern California Law
       Review 46 (June): 617-660.

           Tribe discusses the limits of instrumental, or purpose-
       oriented, rationality.   The most fundamental limit is a
       conception of choice in which objects and alternatives are
       regarded as things selected by an actor. This conception
       ignores ways in which choice alters the characteristics of the
       chooser. An alternative is sketched in which choices are
       viewed as constituents of, rather than products of, human
       character and will.

275
Tribe, Laurence H.
  1972 "Policy science: Analysis or ideology." Philosophy and Public
        Affairs 2 (Fall): 66-110.

        Scientific analysis, Tribe claims, is often passion
        masquerading as reason. This thesis is developed by (1) exam-
        ining ways in which modes of analysis—B/C analysis in particu-
        lar—are founded on commitments to ideological principles, and
        by (2) discussing the patterns of distortions that occur in
        several areas of analysis. Tribe offers several remedies to
        the problems he identifies.

276
Tversky, Amos, and Daniel Kahneman
  1975 "Judgments under uncertainty: Heuristics and biases." In
        Benefit-Cost and Policy Analysis 74, ed. Richard Zeckhauser,
        Arnold A. Harberger, Robert H. Haveman, Laurence E. Lynn, Jr.,
        William A. Niskanen, and Alan Williams, pp. 66-80. Chicago:
        Aldine Publishing Company.

        This article determines certain biases introduced into B/C
        analysis by outcome uncertainties. The authors show that
        subjective judgments of risk rely in large part on principles
        that lead to systematic bias. While this result has negative
        implications for B/C analysis conducted for uncertain outcomes,
        the authors offer an encouraging finding. These biases are
        often predictable and offer the possibility that certain pre-
        ferred methodologies and training techniques will result in
        more reliable B/C findings.

277
U.S. Congress, House Committee on Interstate and Foreign Commerce
  1980 Use of Cost-Benefit Analysis by Regulatory Agencies. Joint
        Hearings, Subcommittees on Oversight and Investigations and on
        Consumer Protection and Finance. Ninety-Sixth Congress, First
        Session (July and October, 1979). Serial No. 96-157.
        Washington, D.C.: U.S. Government Printing Office.

        These joint hearings before the House Subcommittees on
        Oversight and Investigations and on Consumer Protection and
        Finance embody a loosely structured debate between assorted
        proponents and critics of BCA. Among the proponents are James
        Miller III (codirector of the American Enterprise Institute)
        and Murry Weidenbaum (Center for Study of American Business,
        Washington University). Those critical of BCA include Mark
        Green (Public Citizens' Congress Watch), Ralph Nader (Center
        for Responsive Law), and Norman Waitzman (Corporate
        Accountability Project).
        The context for the three-part hearing is the Reagan
        administration's efforts to reprivatize the American economy by
        bridling regulatory initiatives. At one level, testimony is
        focused on controversy over who pays and who benefits from
        federal regulation and in what amount. Its essence is best
        captured in the prepared statements by Green (emphasizing
        regulatory benefits) and by Weidenbaum (emphasizing costs). At

another level, BCA emerges as a fulcrum in the larger debate over how public or how private the American economy should be. References in the prepared statements supply readers with the latest research for and against BCA, broadly defined here to include risk assessment and cost-effectiveness analysis.

278

U.S. Congress, Joint Economic Committee
   1973 Benefit-Cost Analyses of Federal Programs. A Compendium of Papers Submitted to the Subcommittee on Priorities and Economy in Government of the Joint Economic Committee. Ninety-Second Congress, Second Session. Washington, D.C.: U.S. Government Printing Office.

This document begins with a survey of federal evaluation practices and concludes that more extensive and higher-quality analysis is needed. Includes eleven papers by various authors which examine the limitations of conceptualization and practice of B/C analysis as applied to specific projects.

279

U.S. Congress, Joint Economic Committee
   1969 The Analysis and Evaluation of Public Expenditures: The PPB System. A Compendium of Papers Submitted to the Subcommittee on the Economy and Government of the Joint Economic Committee. Ninety-First Congress, First Session. Washington, D.C.: U.S. Government Printing Office.

Volume I contains a number of proposals for improving the analysis of public expenditures, including techniques relevant to distributional issues, risk and uncertainty, multiple objectives, and shadow pricing. Volume II is an extended discussion of the status of the Planning-Programming-Budget System (of which B/C analysis is an integral part) used to evaluate federal investments. Volume III contains articles including several that are critical on the actual performance of the PPB system.

280

U.S. General Accounting Office
   1982 Improved Quality, Adequate Resources, and Consistent Oversight Needed If Regulatory Analysis Is to Help Control Costs of Regulations. Report to the Chairman, Committee on Governmental Affairs, U.S. Senate (November 2). GAO/PAD-83-6.

In response to a request by William V. Roth, Jr., chairman of the Committee on Governmental Affairs of the U.S. Senate, GAO examined ways in which regulatory analyses (RAs), used to judge the desirability of a regulation, can be improved. Though RAs date back to 1974, a recent executive order (No. 12291) requires that "major" regulations be analyzed to assess their costs and benefits.
      The report is a response to six questions raised by Senator Roth for GAO study: 1) How good are the RAs done by the agencies? 2) What are the potential costs of RAs of new and existing regulations, as required by the executive order or

by pending legislation such as S. 1080, and to what extent
could the costs of these analyses distract agencies from their
primary responsibilities?   3) How has the RA requirement
affected deregulatory initiatives, and to what extent will the
proposed legislation interfere with President Reagan's program
to deregulate the economy?   4) What has been the effect of
centralizing regulatory oversight at OMB?   5) To what extent do
the provisions of Executive Order 12291 and S. 1080 conflict
with or preempt existing regulatory legislation?   6) What
effect will presidential oversight of RAs have on independent
regulatory agencies?

In addressing these questions, GAO examined the results of
similar investigations performed previously and interviewed
officials at eleven different regulatory agencies and at OMB.
GAO found that many of the RAs, including several approved by
OMB, do not provide adequate support for their conclusions,
despite an average cost of more than $200,000 for each analy-
sis.   OMB, in its oversight role, has not done all it could to
improve the quality of the RAs or the use of economic analysis
in regulatory decision making.   GAO recommends that OMB oversight be carried out more openly and consistently and that
Congress give greater attention to the resources needed to
prepare good RAs, to the compatibility of substantive legislation, and to presidential oversight under Senate Bill 1080.
(Note: This review relies heavily on GAO's summary of the
document.)

281
Vanags, A. H.
    1975 "A reappraisal of public investment rules."   In Current
          Economic Problems: The Proceedings of the Association of
          University Teachers of Economics, Manchester, 1974, ed. Michael
          Parkin and A. R. Nobay, pp. 119-150.   London: Cambridge
          University Press.

          Vanags develops an intertemporal general equilibrium model
          assuming certainty and maximization of public welfare to
          investigate discount rates.   He finds that, in general, the
          appropriate discount rate is a weighted average of social time
          preference and the marginal productivity of capital in the
          public sector.   Though B/C analysis is not addressed directly,
          the logic and subject matter of the piece are relevant to B/C
          discussions.

282
Van Horn, Andrew, and Richard Wilson
    1976 The Status of Risk-Benefit Analysis.   Cambridge, Mass.: Harvard
          University Energy and Environmental Policy Center.   Available
          from National Technical Information Service as BNL-22282.

          Risk-benefit analysis is a methodology that attempts to
          incorporate the probable cost of low-risk events into a B/C
          framework.   The article reviews the progress made by those
          seeking a refinement of the technique and discusses problems
          that have arisen.   The authors believe that risk-benefit
          analysis should be done in conjunction with the responsible

decision maker so that the values inherent in any decision-
making process can be addressed clearly in the analysis. This
is seldom the way such analyses are carried out.

283
Weisbrod, Burton A.
    1968 "Income redistribution effects and benefit-cost analysis." In
          Problems in Public Expenditure Analysis, ed. Samuel B. Chase,
          pp. 177-209. Washington, D.C.: Brookings Institution.

          Welfare economics--that branch of economics concerned with
          the economics of policy decisions--addresses two broad areas:
          (1) the analysis of efficiency of public programs, and (2) the
          income redistributive effects of programs. Weisbrod contends
          that equity considerations confront economists with difficult
          and controversial value judgments. This has resulted in econo-
          mists concentrating on efficiency while almost totally ignoring
          equity. Weisbrod presents the case for integrating equity into
          B/C analysis and presents two methods for doing this.

284
Weisbrod, Burton A., and Mark Schlesinger
    1981 Benefit-Cost Analysis in the Mental Health Area: Issues and
          Directions for Research. National Institute of Mental Health,
          Series EN, No. 1, Economics and Mental Health. DHHS Publica-
          tion No. (ADM) 81-1114. Washington, D.C.: U.S. Government
          Printing Office. (Also reprinted by Institute for Research on
          Poverty, University of Wisconsin-Madison Reprint Series, No.
          430)

          The emphasis of this article is on the identification of
          problems in the application of BCA to the field of mental
          health. Among the difficulties discussed are: (1) Identifica-
          tion and quantification of the problems is difficult. The lack
          of a program does not necessarily mean that certain things will
          not be done. (2) Consensus on what constitutes mental health,
          or an improvement of it, is problematic. (3) Even with an
          agreed-upon definition, measurement problems abound, especially
          in converting mental health benefits into monetary units. (4)
          The BCA framework may lack the necessary comprehensiveness to
          include all relevant costs and benefits. (5) Budget con-
          straints will frequently limit the scope of the analysis, with
          undesirable consequences.
          The authors point out that these difficulties are not
          unique to this field, but that they are of unusual significance
          in relation to mental health. Decisions on whether the bene-
          fits outweigh costs of a mental health program must be made
          with or without formal analysis. The potential of BCA should
          not be exaggerated; however, the authors claim that BCA can
          inform the judgment process.

285
Williams, Alan
    1973 "Cost-benefit analysis: Bastard science? and/or insidious
          poison in the body politic." In Benefit-Cost and Policy
          Analysis 72, ed. William A. Niskanen, Arnold C. Harberger,

Robert H. Haveman, Ralph Turvey, and Richard Zeckhauser, pp. 48-74. Chicago: Aldine Publishing Company.

Williams argues that B/C analysis is limited in certain well recognized ways. Thus, its proper objective should be to assist rather than dictate the decision-making process. He concludes that, while criticisms of such analysis are valid, BCA is still a valuable--if imperfect--tool.

286
Zeckhauser, Richard, and Elmer Schaefer
  1968   "Public policy and normative economic theory." In The Study of Policy Formation, ed. Raymond A. Bauer and Kenneth J. Gergen, pp. 27-101. New York: The Free Press.

The authors analyze the decision-making process when the problem at hand is complex and involves a large number of interested parties. They conclude that when conflicting interests are present, there is no determinant best solution to a policy problem and that quantitative methods cannot serve as irrefutable arbiters of these conflicts. At best, such methods can simplify the dimensions of the problem at hand.

287
Zimmerman, Burk K.
  1978   "Risk-benefit analysis: The copout of government analysis." Trial 14 (February): 43-47.

According to Zimmerman, the real purpose of risk-benefit (and B/C analysis) is to mask the real risks and costs of activities dangerous to society and to protect the profits (benefits) of those engaged in these activities.

Index

Employment impacts of (continued):
  energy conservation, 038, 070,
    120
  energy crisis, 046
  energy development, 148, 154
  energy pricing, 088, 099
  energy shortages, 155, 160, 171
  environmental control, 077, 079,
    085, 102, 112, 118, 123, 147,
    150, 159, 161, 162
  residential conservation
    programs, 056
  solar energy development, 098,
    104, 136
  wind energy systems, 006
  see also Manpower requirements
Energy conservation:
  consumer choices and, 069, 070
  construction industry and, 074
  distribution of, 126
  employment and, 023, 028, 070,
    120
Energy costs:
  of beef protein and soy protein
    production, 072
  of various goods and services,
    026
Energy crisis, distribution of
  impacts, 128, 137
Energy demand:
  planning for, 119
  projections of, 013
  residential, 082, 128
  substitution studies, 012, 013,
    016, 028, 032, 047, 088, 155
Energy development:
  discount rate for planning, 228
  domestic gas supplies, 002
  manpower requirements of, 106,
    108, 109, 110, 148, 154
Energy-economy models, see Energy
  demand, substitution studies
Energy policy:
  black employment and, 003, 014,
    090
  criteria suggested by Barry
    Commoner, 036
  economic growth and, 089, 170
  economic impact of, 135
  employment effects of, 158, 168
  employment policy and, 156, 163
  NAACP position on, 170
Energy prices:
  distribution of impacts of, 007,
    100, 116, 164
  employment impacts of, 011, 073,
    088, 099

  low-income households and, 095,
    100
  political impact of, 40
  world economy and, 060
  residential, 007, 164
Energy shortages, employment
  impacts of, 160, 171
Energy technology, and economic
  growth, 027
Energy use:
  economic growth and, 094, 149,
    171
  employment and, in Australia,
    005
  family income and, 026, 081, 124
  impact on labor productivity and
    GNP, 132
Environment, amenity values of,
  225
Environmental benefits assessment,
  265
Environmental controls:
  availability of capital and, 048
  economic growth and, 027
  employment impacts of, 039, 079,
    084, 102, 105, 112, 118, 123,
    147, 159, 161, 162
  effects on multinational
    corporations, 166
  household income and, 085
  impact on construction industry,
    150
  job loss and, 017, 221
  labor productivity and, 041, 134
  logging industry and, 169
  mining industry and, 172
  steel industry and, 083, 093,
    162
  see also Environmental programs
Environmental impact assessments,
  see Benefit-cost analysis
  and environmental programs
Environmentalism:
  backlash on, 022
  elitism in, 117, 121, 138, 145
  financial difficulties of, 142
  in the 1980s, 092
  public opinion of, 114
Environmental policy:
  benefit-cost analysis and, 245,
    250, 259
  labor and, 113
  summary of issues, 189
  under different political
    theories, 151
  U. S. economy and, 079, 080, 086

About the Compilers

FREDERICK H. BUTTEL is Associate Professor of Rural Sociology at Cornell University. He is co-author of *The Rural Sociology of the Advanced Societies* and *Environment, Energy and Society* and has authored numerous articles for *Social Science Quarterly*, *Social Forces*, *The American Sociologist*, *Human Ecology*, *Sociological Quarterly*, and *Environment and Behavior*.

CHARLES C. GEISLER is Assistant Professor of Rural Sociology at Cornell University and co-author of *The Community Landtrust Handbook*, *Land Reform, American Style*, and *Indian-SIA: The Social Impact Assessment of Rapid Resource Development on Native Peoples*. He has contributed articles to *Rural Sociology* and *Land Economics*, among others.

IRVING W. WISWALL is Research Support Specialist for Cornell Institutefor Social and Economic Research.

Date Due